2-14-06

To MARV

I KNOW OF YOUR INTREST
IN PCM AND HOPE YOU
WILL FIND THIS BOOK USEFUL

WARMEST REGARDS,

[signature]

PRODUCT LIFECYCLE MANAGEMENT

PRODUCT LIFECYCLE MANAGEMENT

DRIVING THE NEXT GENERATION OF LEAN THINKING

MICHAEL GRIEVES

McGraw-Hill

New York Chicago San Francisco Lisbon
London Madrid Mexico City Milan New Delhi
San Juan Seoul Singapore Sydney Toronto

1 2 3 4 5 6 7 8 9 0   FGR/FGR   0 9 8 7 6 5

ISBN 0-07-145230-3

This publication is designed to provide accurate and authoritative information in regard to the subject matter covered. It is sold with the understanding that the publisher is not engaged in rendering legal, accounting, or other professional service. If legal advice or other expert assistance is required, the services of a competent professional person should be sought.
—*From a declaration of principles jointly adopted by a committee of the American Bar Association and a committee of publishers.*

McGraw-Hill books are available at special quantity discounts to use as premiums and sales promotions, or for use in corporate training programs. For more information, please write to the Director of Special Sales, McGraw-Hill Professional, Two Penn Plaza, New York, NY 10121-2298. Or contact your local bookstore.

 This book is printed on recycled, acid-free paper containing a minimum of 50% recycled, de-inked fiber.

To my wife, Diane,
of five wonderful years

Contents

Acknowledgments

MY FORMAL INTRODUCTION to Product Lifecycle Management, or PLM, came over coffee. I met Gary Baker, a vice president with EDS working on the General Motors account and a longtime colleague of mine from the days when I chaired the Michigan Technology Council. I was planning a session for the Management Briefing Seminar (MBS), which in spite of its inauspicious name was and is an important get-together of the automotive industry organized by the University of Michigan and the Center for Automotive Research each summer in the beautiful resort area of Traverse City, Michigan.

In previous years I had organized sessions at MBS around the topic of information exchanges. Material exchanges were all the rage during the Internet era. Material exchanges in the automotive industry appeared to be win-lose propositions, with the powerful automobile manufacturers certain to wring every last bit of profit from the supplier community.

My perspective on exchanges was a little different. It involved a focus on information and not material. If the automotive community could develop exchanges of information such as order flow, product specifications, product status, warranty costs, etc., then costs could be replaced by this information. This could be a win-win proposition for both the automobile manufacturers and their suppliers. Gary, who had attended my session the previous year, said, "Your ideas about information exchanges are related to something we are working on called Product Lifecycle Management (PLM). Let me introduce you to our people."

The Internet bubble burst, taking with it a good deal of interest in exchanges. However, the interest in PLM continued to grow and the ideas about PLM continued to mature. A lot of my ideas came about because Ed Borbely, the Director of the Center for Professional Development in the College of Engineering at the University of Michigan, encouraged me to develop the first university-based executive education course in PLM. The interaction with executives and managers involved with PLM allowed me to crystallize my thoughts about PLM as a larger approach to the product information management problem and its relationship to other approaches and systems.

I would also like to thank my professional friends and colleagues who have helped me refine my ideas about PLM: John Crary, the CIO of Lear Corporation who was my partner in crime in developing information exchanges and implemented PLM, which gave me a window into its actual use; Lorie Buckingham, CIO of Visteon Corporation, who helped me refine some ideas on the strategy of moving a large organization into PLM; Eric Sterling, VP UGS; Peter Schmitt, VP Delmia; Raj Khosho, VP UGS, who helped me clarify the use of information as a substitute for wasted time, energy, and material at a workshop in Qingdao, China; Ed Miller, CEO of CIMdata, who has presented an overview of the ever-evolving PLM supplier community in my courses; and Nino DiCosmo, Chairman and CEO of Autoweb, Inc, who had planned a peaceful trip with me to Tokyo for a board meeting, but spent it having to critique the issue of information singularity.

My editor, Jeanne Glasser, deserves a great deal of credit for identifying PLM as a topic worth reading about, tracking me down, and convincing me that writing a book would be fun—and then having to put up with the problems of a new author.

Last, but not certainly not least, I would like to thank my lovely wife, Diane, for encouraging me to write this book and then leaving me alone to do it. My son, Rob, and his wife, Chris, were also encouraging. Their children, Nick, Bianca, Jake, Gabrielle, and Bella constantly remind me by their observations and actions that there are new, exciting, and useful ways of looking at the world, even if, as Nick who is 12 puts it: they "have no clue" what I'm talking about when I talk about PLM.

Introduction—
The Path to PLM

P RODUCTIVITY IS driven in waves. We create new ways of doing things or new things to do that drive a new wave of productivity. Some waves of productivity are driven by a seminal invention such as the steam engine, the automobile, or the computer. Other waves are driven by our approach to the way we do things, such as the assembly line, the multidivisional or M-form corporation, or lean manufacturing.

As the newest wave in productivity, Product Lifecycle Management—popularly referred to as PLM—emerged in the last few years fully formed, or so it seemed. PLM was first piloted in the automotive and aerospace industries: two sectors with complex, manufactured products. The electronics industry, which has product management issues that focus more on software configuration than the complex product configurations of the automotive and aerospace industries, was also an early adopter of PLM or PLM-like technologies. With the success of PLM in these three industries, interest in PLM has spread to businesses as diverse as consumer packaged goods (CPG), industrial goods, medical devices, and even pharmaceuticals.

PLM is an outcome of lean thinking—a continuation of the philosophy that produced lean manufacturing. However, unlike lean manufacturing, PLM eliminates waste and inefficiency across all aspects of a product's life, not solely in its manufacture. PLM is focused on using the power of information and computers to deliberately pare inefficiencies from the design, manufacture, support, and ultimate disposal of a product. Wherever possible, PLM enables the movement of inexpensive information bits in place of expensive physical atoms, a concept popularized by Nicholas Negroponte.[1]

In doing this, PLM takes "lean" to the next level. Lean manufacturing is a continual process that works at taking out the inefficiencies in the manufacturing process. However, as lean manufacturing efforts find and eliminate waste, products are being produced less efficiently at other phases of development. PLM uses product information, computers, software, and simulations to produce the first product as efficiently and as productively as the last product throughout the design, development, and delivery process.

Lean manufacturing requires considerable resources because changes that improve production cause equipment to be reconfigured, machines rearranged, and material relocated as the lean manufacturing engineers test their hypothesis that this new method will decrease waste. Once the system is set in place, PLM uses little in the way of resources, since this same process is done digitally.

Testing lean approaches is time intensive, so only the most promising ideas for streamlining the manufacturing process can be tried. The wall clock ticks away as the new configurations are set up, production commences, and the results are evaluated. PLM does not operate under the same time constraint. PLM can simulate wall clock time, and it can do multiple versions of it simultaneously, so all hypotheses can be tested, not just the most promising.

Finally, lean manufacturing can only take an organization so far. The most efficiently produced product resulting from the best lean manufacturing processes can be flawed as a result of design failure or failure in actual use. It is nothing more than efficiently produced scrap that is a waste of time, energy, and material. Productivity increases in the production of scrap are a disappointing, but logical, result of a limited approach to lean manufacturing.

Lean Thinking–Globally!

Seeing what lean thinking can do on the manufacturing floor has left companies eager to extend these benefits of lean into other parts of the organization. But, to do so, lean will have to be accompanied by an integrated approach to product information and the tools and techniques needed to enable that integrated approach. The level of productivity that PLM can drive promises to be enormous, as evidenced by the attention it has received in a short period of time.

PLM has attracted worldwide attention on a global basis; it is not solely an American or European initiative. It is being adopted by organizations everywhere. We expect organizations based in the more industrial Asian countries such as Japan and Korea to be early adopters of PLM. However, organizations in such diverse countries as India, Malaysia, and China are also not only adopters, but innovators of PLM.

PLM is able to raise the bar on productivity because it allows for the complete integration of everything related to a product or service—both internal and external—into the organization producing it. As you'll learn as you read this book, PLM uses information technology and organizational practices and processes to improve efficiencies both within and across functional areas. Dividing work along functional areas, such as engineering, manufacturing, sales, and service, is an organization's method of dividing tasks in order to simplify complexity.

In the past, a great deal of effort and focus has been placed on increasing efficiencies within these functional areas. Although improvements can always be made within the various functional areas, these initiatives suffer from the law of diminishing returns. The high-return projects have been identified and remediated. This is especially true of those companies that have embraced Six Sigma project teams, where their mantra is continual improvement.[2]

In fact, PLM initiatives are becoming an option for Six Sigma teams looking for areas of improvement. Because PLM generally originates in a specific departmental area, it may be natural simply to view PLM as a functional area initiative. PLM projects can naturally start in engineering, because that is where product information originates, and there are a substantial number of opportunities to make improvements through better organization of product information. However, as we shall see throughout this book, the bigger opportunity is to use PLM to enable better information flow across the entire organization.

Functional areas can easily become isolated silos, with little communication or coordination among them. Attempts to optimize performance within these silos can actually lead to substantial underperformance across the whole organization and its related supply chain.

PLM holds the promise of improving productivity through a cross-functional approach, using product information. By linking different functional areas through shared product information, PLM can help organizations break down the silo perspective and unlock productivity gains as functional areas benefit from a shared base of information. As supply chains become more integrated, PLM has the potential for impact across these supply chains—not just within the organization. This will enable productivity and performance gains that cannot be obtained if the focus is solely on individual areas.

The other allure of PLM is that it does not improve efficiency and productivity from simply a cost-reduction perspective, but also from a revenue perspective. Increasing costs are not an inherently bad thing. If revenues are increasing, it is almost impossible not to increase costs. The key to increasing profits is just not to let costs increase at a faster rate than revenues.

PLM has within its framework the opportunity to increase innovation, functionality, and quality—three drivers of increased revenues—by better organizing and utilizing the intellectual capital of an organization. The ability to develop and build creative, more useful, and better products from the same amount of effort will also drive productivity and is a great deal more sustaining than cost cutting. As the old adage goes, "You can't simply save your way to prosperity." Real prosperity requires revenue growth.

At first blush, PLM appears to be a relatively straightforward concept. As the name implies, it is the management of the information about a product throughout its entire life cycle from initial design to final disposal. However, as will be explained in the next chapter, the devil is in the details of this seemingly obvious explanation. In addition, there is still a good deal of discussion and disagreement regarding the form, scale, scope, and implementation of PLM.

Even in its initial phases, PLM is a "big idea" information technology undertaking. Similar to Enterprise Resource Planning (ERP) initiatives, PLM's greatest promise is not in the foundation projects that affect one functional area, but in its larger strategic use that is cross-functional, enterprise-wide, or even supply chain inclusive.[3]

However, the days of the chief executive officer (CEO) and chief information officer (CIO) going to their board of directors and saying, "Give us $500 million and two years, and we'll give you an enterprise system" have come and gone—if they ever really existed. So too are the days when any project involving the Internet received automatic approval without the annoyance of having a financial justification or even a business proposition that was quasi-logical.

As John Crary, CIO of Lear Corporation, a $13 billion automotive supplier says, "The only way CIOs will bring projects to their board for approval is if they have a well defined Return on Investment (ROI)."[4] PLM holds that promise, and it does it on a "pay as you go" basis, as we shall see. This is the only way that such a broad technological concept could even hope to be funded in this day and age.

In this introduction, we will explore why PLM and other information systems have the potential for such a powerful impact on the productivity of an organization. This will be the basis for the claim that PLM will drive the next wave of lean thinking in organizations that adopt and embrace PLM. The success of PLM relies on some underlying fundamental premises. We will explore four of them in this introduction. We use these premises every day to guide our decisions regarding information technology adoption. However, we often do not realize it. Nor do we realize the increasing impact of these premises. The premises we will discuss are: information as a substitute for time, energy, and material; the trajectory of computer technology development; the virtualization of physical objects; and the distinction between processes and practices.

Information as a Substitute for Wasted Time, Energy, and Material[5]

We often lose sight of why Information Systems (IS) have such a powerful impact on organizations. Systems that enable approaches such as PLM are developed and find a place in organizations for a fundamental reason: with these systems, we can substitute the use of information for the inefficient use of time, energy, and material. Since we live in a physical world, we cannot substitute information for all uses of time, energy, and material, only for those used inefficiently. With physical products, we eventually have to do something with atoms: move them, shape them, reconfigure them, assemble them, etc. To produce physical products, we have to use material, expend energy, and use people to do so.

Let's take an example from everyday life to illustrate this principle of substituting information for wasting time, energy, and material. Those of us who play golf will be all too familiar with this example. We hit our first golf shot off the tee. We then drive our golf carts up to where the golf ball rests. Because we need to know the distance to the green in order to select the right club and hit our next shot accurately, we need to find out where we are. While there are yardage markers on the course marking the distance to the green, we usually have to find them. This entails some time and some energy to drive our golf cart around to find these markers.

We then compute the distance between these yardage markers and our ball, sometimes by stepping off the distance between the yardage marker and our ball—again taking more time and more energy. We also sometimes drive our golf carts farther up the course to survey the green to see if there is water or other nasty hazards that might come into play on our next golf shot. Again, this entails more time and energy. Only after we use this time and energy, do we select our club and hit our next shot.

Contrast this with golf carts that are equipped with a small computer and GPS system that show us an image of the hole, where we are on that hole, and yardage to the green and to hazards that we ought to avoid. We now drive up to our tee shot, look up at the computer screen, and know precisely how much yardage we have to the green and what hazards we ought to avoid surround the green. We can immediately get out of the cart, select the right club, and hit our shot.

While the waste of time and energy on the golf course is unlikely to become a national issue, it does illustrate nicely the substitution of information for the waste of time and energy in a common, familiar task. While we may have the luxury of this waste in a leisure task situation, the situation is very different when our objective is to minimize the use of resources, as we need to do in for-profit organizations.

Time, energy, material, and information are not directly comparable because, with respect to quantities, they all have different units of measure.[6] However, with respect to the value we typically place on them, we can compare them because we can translate each of them into a cost, and then compare the costs.[7]

Figure 1.1 represents the relative costs of a typical task that we perform. It could be designing a product that has specific functions, drilling a hole in a part that will match up with another part to be bolted on, routing material through three stations on a production floor, or, on a personal level, assembling a bicycle to put under the Christmas tree. On the left bar, there are three components of that task. The lower part of the bar represents the cost function of the time, energy, and material we would expend if we did the task in the most efficient manner possible. We waste no material. The time to perform the task is the least possible time of all possible ways to perform the task. We minimize the amount of energy used. With today's focus on "lean" manufacturing and other functions, this is the optimal "lean" task.

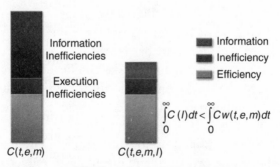

Figure 1.1   Information as Time, Energy, Material Trade-Off

The second or middle part of the bar is execution inefficiencies. These are inefficiencies that develop because, even though we know what the optimal procedure is to perform the task, we just do not do it properly. We design the part, but forget to include one of the functions. So we have to redesign the part to include the forgotten function. We drill the hole in the wrong place when producing the product, and we have to scrap the material. We also waste the time and energy we have used to drill the misplaced hole. We move the wrong material to the wrong station on the factory floor, and we have to relocate it to the right station, wasting time and energy. We select the wrong bolt to put the wheel on the bicycle, and we have to dissemble the bike and reassemble it with the right bolt.

The top part of the bar represents the information inefficiencies. This is the inefficient or wasted use of time, energy, and material because we just do not have the information required to do the task efficiently. We do not know how to get all the functions incorporated into a single design without trial and error. We do not know exactly where the hole to be drilled is or what tolerances we need so that the hole lines up with the other part later on in the production cycle. We do not know what the machine loading is on the factory floor so that we can route the material to the machines with capacity. And something we have all experienced, we just cannot tell from the bicycle plans—supposedly written so a child could understand them—what size and type of bolt among all the bolts in the package is the one required to hold the front wheel on the bicycle.

On the right bar, we have the relative impact on costs of information replacing inefficient or wasted time, energy, and material. The bottom part of the right bar is still the same optimal use of time, energy, and material for our task. The middle part of the right bar, execution inefficiencies, is still the same. While an area of concern, remedying execution inefficiencies is where engineering usually plays its part. Engineering excels at taking task inefficiencies where the optimal use of time, energy, and materials are known, and devising processes and machines to reduce those inefficiencies as much as is possible in an imperfect physical world. Six Sigma project teams are aimed at these execution inefficiencies.

The top part of the right bar is where information can replace the inefficient or wasted use of time, energy, and material on the top part of the left bar. Our vault of drawings shows us how the functionality we need was accomplished in previous designs. Computer Aided Design (CAD) and engineering specifications make it possible to know exactly where to drill the hole so it will fit perfectly every time with another part, with no wasted material or the time and energy necessary to process that material.

Information systems that monitor each work station let us know which machines are available at what specific times so that the material moves continually through the three machines in the least amount of time with the least amount of effort. An instructional video showing someone assembling the bicycle with a close-up on the parts used at each step minimizes the wasted effort dismantling and reassembling the bicycle by trial and error. (Although, from experience, the author would recommend the most efficient use of time, energy, and material would be to have the store that assembles hundreds of these bicycles do the assembly.)

Figure 1.1 shows a substantial reduction in overall cost resulting from the substitution of information for the inefficient use of time, energy, and material. However, Figure 1.1 is only meant to be an illustration and does not reflect the actual cost function for any particular task. Admittedly, it is also an ideal representation.

As shown in the right bar, information might not replace all the inefficient use of time, energy, and material. In a typical situation, information might substitute for a substantial amount of wasted time, energy, and material, but not all of it. For complex tasks, the most efficient use of time, energy, and material might not be knowable because the permutations and combinations are so vast.

For simple tasks, the inefficient use of time energy, and material might be very low. Informational costs might not be lower than the inefficient use of time, energy, and material. In fact, informational costs may be higher under those circumstances. Also, for tasks that were only performed once, it might be more cost effective to use a trial-and-error approach than to incur the costs of collecting and processing the required information.

For more complex, repetitive tasks, the cost of information may be substantially less than the inefficient use of time, energy, and material. Informational costs might be a fraction of the cost of the inefficient use of time, energy, and material. In spite of the lack of general quantitative models that confirm this relationship, our behavior in investing in information systems, both formal and informal, would indicate that we have a fundamental belief in this relationship between inefficient use of time, energy, and material and its substitution with information. This is the underlying assumption and basis for all information systems, and especially for those that implement PLM.

We have long recognized this in the manufacturing function. Lean manufacturing substitutes the information about a more efficient way to do a process for the wasted time, energy, and material of the less efficient process. It may be that manufacturing is more deterministic and can be better assessed for this waste, but other functions also waste time, energy, and material.

The estimates of waste in the design and engineering function are often cited at 60 percent to 80 percent of the total design and engineering costs! These wastes are analogous to the readily recognizable manufacturing production wastes, such as wasted motion (engineers waiting around for meetings), scrap (developing parts with design specifications and drawings that are outdated), overproduction (designing parts that have already been designed before), rework (designing parts that cannot be manufactured and must be redesigned), material shortages (looking for drawings and engineering data), and transportation waste (copying/moving drawings and engineering data). This is not to pick on engineering. There are analogous wastes in all phases of the product's lifecycle that can benefit from the substitution of information for wasted time, energy, and material.

There is one misconception about information that often is taken for granted. While it is cheaper to move informational bits than physical atoms, information is neither costless[8] nor, as any organization that examines its IS budget knows, especially cheap. This is very different from the assumption that the information needed to replace the time, energy, and material used inefficiently is less expensive than simply wasting that time, energy, and material.

Put another way, the assumption underlying the substitution of information for wasted time, energy, and material is that the sum of the cost of this information over the life of the tasks is less than the sum of the inefficient use of time, energy, and material over that same lifetime. The formula that reflects that assumption is in the right lower corner of Figure 1.1.

Under normal conditions, especially where tasks are repeatable, this is not an unreasonable assumption. Learning and experience curves, where production costs decrease with the number of units produced, are examples wherein information about repeatable tasks replaces wasted time, energy, and material. Figure 1.2 is an example of an 80 percent learning curve. For every doubling of production units (1X, 2X, 3X, etc.), costs decrease by 20 percent. This is the learning or experience curve that is normally thought to occur naturally in organizations, although other organizations have focused on obtaining steeper curves.

The idea that costs decline as units of production increase is an intuitive one that goes back to the early workshops of civilization. Its modern articulation is credited to T.P. Wright's observations in 1936 regarding the unit cost of airplane production falling as the number of units increased.[9] The general idea is that as organizations gain experience (i.e., information) with producing and manufacturing a product, their unit production costs generally decrease over time. These cost decreases are not simply because of information learned by the production workers who are actually producing the product. Other areas of the organization, such as engineering, purchasing, and administration, also participate in generating the cost decreases. Even organizations where the manufacturing or

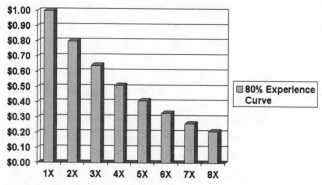

Figure 1.2 Experience Curves

production process is highly automated so that learning from the manufacturing or production process is substantially limited still show cost decreases with cumulative unit production.

The driver for this cost decrease is not simply the doubling of production. Rather, it is the feedback loop as management and workers perform their production tasks, analyze their actions, make adjustments and corrections, and start the cycle again. As management and workers develop information based on their experience in producing the products, they are able to substitute that information for the inefficient use of time, energy, and material.[10]

However, a fair amount of time, energy, and material is wasted in traveling from the upper left-hand corner of the cost curve to the bottom lower right. PLM and other information systems actively attempt to substitute information from the beginning of the cost curve and capture the wasted time, energy, and material much sooner than normally would be the case.

Looking at the cost decrease differently, a larger number of units can be made at the lower right of the curve than can be made at the upper left of the curve for the same amount of cost. This increase in units for the same amount of cost represents a productivity increase. Historically these experience cost curves have been examined for the production facility.

But there is no reason that these cost curves cannot be looked at across the entire organization, across the supply chain, and across the product's entire life cycle. If information about the product's design, manufacture, use, and disposal can be captured and substituted for the inefficient and wasted time, energy, and material across the product's life cycle, then this is a new source of productivity—one that extends beyond the functional area of production and manufacturing and extends to the entire organization.

The Trajectory of Computer Technology Development

The story of the trajectory of computer technology development can best be told in an old tale of a king and a wise man. The tale comes from India and tells of a king who, delighted by the advice of a wise man, asked him to name his reward. The wise man demurred because he had not performed the service for the king in the hope of reaping a reward. As kings are wont to do, the king did not take no for an answer and commanded the wise man to name his reward. The wise man pointed to a chess board next to the king and told him that he would like to start with one grain of wheat and double it each day for each square on the chessboard. The king laughed at such a seemingly meager reward and waved his hand to grant it.

The king made a show of having the first grain of wheat delivered by a servant bearing it on a silver tray. The king promptly forgot about it, but not for long. By the sixteenth day, the amount had become a pound of wheat. By the twenty-sixth day it had become 1,000 pounds of wheat.

The king computed that by the thirty-second day he would need to send the wise man slightly over 131,000 pounds of wheat. The king quickly realized that there would not be enough wheat in his kingdom, or even in the entire civilized world, to fulfill his commitment. He was forced to summon the wise man and ask to be released from his commitment. The wise man had made his point about wishing not to be rewarded.

The king could not foresee the impact that this doubling would have on his kingdom. In a much more positive frame, we also have not foreseen the impact that the doubling of computing capability would have on us, although we have been quick to take advantage of it. Over the years, we have looked for increasingly complex applications to take advantage of the rapid doubling of capability. PLM is our latest effort to develop capabilities that require increasing computing and communications capabilities.

The reality is that PLM could not develop on the scale and scope envisioned without the advancement, both past and forecasted, of computing and communications infrastructure technologies: computing power, storage capacity, connectivity, and bandwidth. PLM applications today require that technology capabilities reach a level that will support its robust requirements. As will be shown later, PLM may be the most Information Technology (IT) resource-intensive general purpose application.[11]

The math-based representations of PLM product information require substantial computing capabilities. The computing capability required to render visual representations of the math-based data and manipulate those representations is substantial. Previously, these capabilities were only available in high-end, expensive graphics workstations.

However, these capabilities have migrated to standard personal computers as these computers continually increased in power and capacity. Once that occurred, around the end of the twentieth century, the use of math-based product descriptions became feasible for use by a wider population and for a wider variety of functions.

This continual increase in power and capacity is the result of the increases in computing power predicted by Moore's Law. Moore's Law is really an observation by Gordon Moore, a founder of Intel Corporation, made almost 40 years ago. Moore's observation was that the computing capability of the microprocessor would double every 18 months. Computing capability has followed this inexorable increase ever since. Since 1971, the number of transistors on a processor increased from 2,250 on the first Intel microprocessor, called the 4004, to 1.7 *billion* transistors on the new Intel Montecito processor scheduled for release in 2005. We are only 20 squares into the chessboard!

In addition, there are corresponding increases in storage and bandwidth.[12] In fact, storage increases are occurring faster than processor chip density increases. Storage densities are thought to double every 12 months. This is due to both technological and production advances and breakthroughs. Bandwidth is on a similar trajectory. In the late 1960s, 110 bits-per-second (10 characters-per-second) on teletype machines was the fastest interactive speed available to users. Users now need and expect megabit speeds and need higher speeds for moving CAD and video files.

The other major technology to enable the development of PLM is the Internet, both directly and indirectly. On a direct basis, the Internet enabled ubiquitous access for almost anyone from almost anywhere. While large organizations had data communications networks, they often were functionally organized, and giving access to outside functions such as manufacturing to engineering was often costly and cumbersome. The Internet changed that as even these organizations adopted Internet standards in place of proprietary ones.

On an indirect basis, the Internet drove technology developers to increase the speeds and capacities of communication channels, which benefited all users. Internet capacity also began doubling on an annual basis.

But the story of computer technology development is not simply about feeds and speeds. It is also about the continually increasing richness of the computing environment that software provides. We are constantly building on the capabilities that currently exist. As we do so, we are also continually enriching the environment we have to work with.

More and more, we are distancing ourselves from having to adapt ourselves to the environment of the computer. The computer is increasingly adapting itself to our environment. When we first began to work with computers, we had to translate anything we wanted computers to do for us into a string of zeros and ones. We had to "talk" their native language, a laborious and cumbersome process.[13]

A few decades later, the early personal computers still demanded substantial adaptation by requiring us to type cryptic commands to invoke special purpose programs. It was only in the mid 1990s that graphical user interfaces (GUIs) masked the machine commands and the computer started to interface itself with us in a form more familiar to us: an image of the desktop, a spreadsheet, the pages of a document. Since then, graphical representations keep getting better and better, until some of the images presented to us are indistinguishable from actual photographs.

Increasingly, computers are dealing with other modes of media, such as sounds and movies. Computers not only passively capture and store these modes, but actively create and manipulate sounds and images in a form and richness that is most comfortable to us. It is not only a matter of comfort and familiarity. It is also a matter of efficiency. We human beings have limited personal input–output bandwidth. Absent a major breakthrough in person–machine interfaces, computers will need to expand their utilization of our highest input bandwidth—our sense of vision.

It is this trajectory of computer development that enables the initial applications of PLM. However, PLM's more ambitious capabilities will require that this trajectory continue unabated, as seems to be the case. While we are more astute than the king in that we have at least calculated the progression of numbers brought about by doubling, we do not know the capabilities that will be enabled by this progression. If the past is any prediction of the future, it promises to be dramatic.

The Virtualization of Physical Objects

Virtualization is not as futuristic as it sounds. All of us can "virtualize" physical objects. If we have a blue coffee cup before us, we can all close our eyes and imagine this blue coffee cup. We take that image and rotate the coffee cup 90 degrees around its horizontal axis and "look" at the coffee cup's bottom. We can modify the image by wrapping a thin, gold band around the top. We can even do simulations by thinking about this coffee cup sliding along a table, off the edge, and onto the floor.

There are however two limitations that constrain the usefulness of this virtualization. The first is that, for most of us, we are limited to simple objects. We can do it for a coffee cup. We cannot do it for it for all the parts of an automobile. The second is that this virtualization cannot be shared with anyone else. This virtualization occurs in our individual, unshareable minds. We can try and describe the image we would like others to virtualize, but we have little confidence in our images being the same as theirs. In order to have everyone have the same image of a physical object, we need to have some sort of physical representation for all to see.

If an early ancestor wanted to show fellow cave dwellers a physical object, for example, a new type of club, then this person needed to have that club in hand. The best this ancestor could do without the club in hand was to use words to describe it to the audience. This was a slow, cumbersome process, prone to errors because it relied on both accurate description and the audience's capability to successfully interpret this description.

We got a little better as time progressed. We learned both to capture the visual features through models and drawings and to abstract information, such as length, width, depth, and weight, which could be described numerically. We did this so we could create models and mock-ups of physical objects in an attempt to virtualize them. We could make elaborate drawings that captured substantial detail about the physical object. In the mid 1800s, we could even combine drawings with the abstracted numerical information to make plans that others could use to replicate the physical objects with almost complete accuracy.

There were a number of trade-offs. One tradeoff, as pointed out by Evans and Wurster,[14] was between richness and reach. Extending the distance and audience required simplifying the representation of the physical object. A scale model could be seen by only those who were physically close in distance. Duplicating the scale model took resources, and physically distributing it across a large geographical area took more resources. Drawings of the physical object could be made with fewer resources than the scale model, and carrying it across distances required fewer resources than carrying the scale model. A simple description could be quickly duplicated and distributed more easily than drawings.

The other trade-off was between resources and richness. As pointed out above, it took more resources to duplicate a scale model than a drawing. It took more resources to duplicate a drawing than to duplicate the descriptions. In the automotive industry, a great many drawings were made before a clay model was done, because it took substantially more time and resources to make a clay model than a drawing.

Another aspect of this is that the representations were static, not dynamic. The representation of the physical object was at a specified point in time. If we desired a representation as the physical object changed, then we had to make another drawing or, at a minimum, alter the drawing of the first representation.

Computers changed this. We could now represent physical objects in a medium that was dynamic rather than static. This allowed us to "virtualize" the representation of the physical object, change it, and easily show the representation of the changed object. At the beginning, the physical object had to be highly abstracted. In the 1970s, two Ping-Pong paddles and a Ping-Pong ball were virtualized on a computer. The two Ping-Pong paddles were represented by two vertical lines and the Ping-Pong ball was a circle of light. However, it was exciting because it was an example of the power of the computer to represent a dynamic, physical activity—a Ping-Pong game.

Since then we have advanced rapidly in the ensuing 30 years. We now have three-dimensional representations of objects that carry with them the information of the physical objects themselves. Not only do we have this three-dimensional image, we have the information about its dimensions, structure, and even material makeup.

So how do we decide that we have virtualized a physical object? I would propose that we decide it by a test similar to that developed by Alan Turing, the famous mathematician and pioneer of computer science. To answer the question of when a computer would be considered intelligent, Turing proposed the Turing Test, an "imitation game," in a famous article in 1950 written for the Oxford University Press philosophical journal, *Mind*.[15]

To answer the question, "Can machines think?" Turing proposed an "imitation game." He proposed that an interrogator send typewritten questions to two subjects, a human being and a computer, from which the interrogator is physically separated and which he cannot see or hear. If the interrogator cannot determine which one of them is human and which one of them is a computer from the typewritten responses of the two subjects to his questions, then the computer is said to "think."

With all due respect to Dr. Turing, who was brilliant but did not have the benefit of our experience with and understanding of computers, he had the wrong element in focus. He was focused on human beings in his imitation game. It might have been more fruitful to have focused on everything but the human being. No computer has yet been able to pass the Turing Test and imitate a human being with its ability to think. However, computers may well be able to imitate the environment that human beings inhabit. In other words, computers may well be able to imitate everything in the physical world *except* human beings.

I would propose the "Grieves Test" for computer virtualization. The general version of the test is to take the same human interrogator and isolate him or her in a room with only a closed-circuit television image of a specified physical object. If the interrogator cannot tell the difference between the actual physical object and the computer representation, then the physical object is said to be virtualized.

Now I would like to propose a visual and performance version of the test. In the visual version, the interrogator can ask to have the objects rotated to any angle. He or she can ask to have the object disassembled and examine the individual components of the object from any view or any angle. He or she can ask to have areas of the object magnified. He or she can ask to have dimensional measurements taken. If the interrogator cannot tell the actual physical object from the virtual one, then the object is said to be visually virtualized or has passed the Grieves Visual Test.

The performance version of the test is more demanding. The interrogator has the same two views of the object, but this time he or she can ask to have any physical test performed on both objects. In addition to simple manipulations and dimensions, they can be tested for weight, durability, aerodynamics, etc. They can be cut open, have power applied to them, and be crash tested.

In short, anything can be done with the objects except to physically picked them up and handle them. However, that too would be possible if the interrogator were suitably equipped with gloves and goggles so as to eliminate the contextual cues that would allow him or her to determine the real from the virtual. If the interrogator cannot tell the actual physical object from the virtual one, then the object is said to be virtual in performance and meet the Grieves Performance Test.

The performance case is more demanding because, not only does the object itself have to be virtualized, but the environment it exists in also has to be virtualized. The performance case requires substantially more computing capability than the visual case, but obviously has more usefulness, especially as it relates to simulation of outcomes.

We will make use of this test throughout this book as we show how PLM substitutes information for the inefficient use of time, energy, and material. PLM can do this only if it can faithfully imitate the actions of the physical world. If it can, then the cost of manipulating bits will always be cheaper and faster than manipulating atoms. PLM's hope of driving productivity will be realized.

The Distinction between Processes and Practices

We like to think that what we do in our organizations is process. Under system theory, process is a deterministic way of linking inputs to outputs. In a systems view of the world, we have inputs, processes, and outputs. For any given set of inputs, we get a well-specified and consistent set of outputs. It is all very neat and well defined.

So we like to talk about processes. Everything in our organization is a process. We have engineering processes. We have purchasing processes. We have manufacturing processes. We have order to delivery (OTD) processes. We have sales processes. If we have all these processes, why do we have such problems in getting consistent outputs from the same set of inputs?

The reality is that we do not have only processes, as much as we would like to think we do. The reality is that we have a spectrum. As shown in Figure 1.3, at the right end of the spectrum is this process view. The process view requires that we have well-defined inputs, so that we can apply deterministic processes and get specific, consistent outputs. The hallmark of a process is that it can be fully scripted or coded.

If we put a piece of material of a specific quality into machine A, run machine A at a certain speed and duration, then we get a part that always meets certain specifications. If we give purchasing a part number and a quantity by 5:00 today, then by tomorrow at 5:00, purchasing will give us a vendor number and an estimated delivery date. These are processes.

At the left end of the spectrum is art. We have fuzzy inputs. We make substantial judgments about the inputs and we cannot explain very well why we do the things with the inputs that we do. We

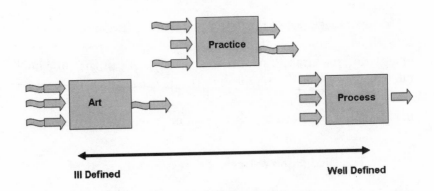

System Approach—Linking Inputs to Outputs

Figure 1.3   Process, Practice, Art Continuum

develop a lot of possible outputs and winnow those down by judgments about how well they match the desired goals of the system. We select outputs that we decide are acceptable, although there may not be a consensual agreement by the receivers of the output that they are of the quality and quantity expected.

Developing an advertising campaign is an art. We have a lot of different inputs that we consider. We develop a number of possible advertising forms and winnow those down. When we select the one we want, it may not produce the desired effect. There are other examples of art within an organization. On the manufacturing side, some forms of casting and grinding are art forms. On the sales side, large, complex sales surely are art.

Organizations usually try to minimize art. Art by definition is not well understood. Art is inefficient. Art leads to inconsistent results and a great deal of uncertainty. Unless an organization is in the business of art, such as moviemaking, organizations try to move art down the spectrum toward processes.[16]

Between art and processes, we have one of the most common forms of an organization: practices. With practices, the inputs are pretty well defined, as are the outputs. What is not well defined is how the inputs are derived from the outputs. There is a substantial amount of judgment and experience of past actions that go into selecting and transforming inputs to outputs. In addition, the outputs may be socially constructed. More simply put, it may take people's agreement to determine if the outputs are acceptable, especially if alternate outputs appear to be acceptable.

In addition, processes take place in reasonably controlled environments. Practices do not. With practices, the context in which they take place adds to the complexity and needs to be considered. Two similar situations might be dealt with very differently in different contexts. For example, a doctor might diagnose the symptoms of a patient very differently if the doctor knew the patient had recently traveled to the Amazon rather than if the patient had never left Iowa all of his or her life.

The classification of engineering computer-aided design (CAD) representations is an example of a practice within an engineering organization. Unless the CAD representation is extremely simple, the engineer classifying it has latitude in how he or she classifies the

CAD representation. Previous experience and the shared understanding of the engineering group will influence how the CAD representation is categorized. Also, the engineer might classify the part differently for a heavy-duty truck than for a high performance sports car. These different vehicles might emphasize different operating characteristics for categorization: one for characteristics of durability, the other for characteristics of performance.

Since this is a spectrum, there are varying degrees of practices. Some activities start out as practices and evolve into processes as experience is gained and agreement reached on all the permutations of inputs that are aligned with outputs. Some manufacturing processes were once practices, as when the shop supervisor altered routings on the factory floor to obtain better throughput. As factory physics became better understood, the routings were generated and controlled by computer, thereby turning a practice into a process.

In general, even when we recognize that our organizations have practices, we exhibit a great deal of misunderstanding and confusion about the distinction between process and practices. When we talk about practices, we usually really want them to be processes, because processes are much neater and better defined than practices.

The most visible example of this is when we talk about "Best Practices." It is as if we can examine the whole range of possible practices over a universe of different contexts, different value judgments, and different consensus outcomes. While we may be able to do that for processes, where inputs and outputs are well defined, we cannot do that for practices.

Judgment of what inputs to consider and how to weight them, along with the context in which we are considering the inputs, make the idea of best practices pretty remote. The training regime that Lance Armstrong, the seven-time Tour de France cyclist, uses may indeed be the "best practice." It would, however, probably kill the average bicycle rider.

The reality of practices is that we really need "Healthy Practices," not "Best Practices." Healthy Practices are those practices that are effective in a wide variety of situations and contexts and will bring satisfactory, not optimal, results. Healthy Practices are attainable by most organizations. At worst, Healthy Practices will not harm the organization that subscribes to them. An example of a Healthy Practice is "Cost engineering changes orders before you agree to them."

The term *Best Practices*, as commonly used today, consists of three distinct categories. The first is "Best Processes," which are not really practices at all, but are very highly defined processes. As stated before, processes have well defined inputs, outputs, and methods by which inputs turn into outputs. In dealing with very well defined processes, we can optimize or come close to optimizing performance based on specific criteria. Machine routing can be a best process.

The second category of Best Practices is "Ideal Practices." Like Lance Armstrong's training regime, they define the ideal practice under ideal circumstances. "Freezing the design early in the design cycle" is an example of an Ideal Practice. It is the ideal thing to do, but most organizations cannot do it because it can only be done under ideal circumstances.

The third category of Best Practices is the idea of "Healthy Practice." As stated before, these are practices that are effective in a wide variety of situations, circumstances, and contexts. Healthy practices will lead to "satisficing,"[17] not optimization. With respect to human practice, we talk about optimizing but generally settle for satisficing. This is not a shortcoming. With human practices, if our standard is optimization, there is a high degree of probability that we will never achieve it. What we can hope to achieve is to accomplish a satisfactory level of performance, which, with fuzzy inputs and outputs, is still an acceptable achievement. When talking about practices in this book, we will more often than not use *satisficing* and not *optimizing*.

In all fairness, a large number of the Best Practices that are promoted are really Healthy Practices. In this book, we will try and focus on Healthy Practices. For material from other sources, we will put "Best Practices" in quotes to indicate that we are not taking them at face value. We will attempt to identify those "ideal" practices that are labeled as "Best Practices."

Why is the distinction between process and practice important, other than to clarify a fuzzy area that leads to confusion? First, it explains some of the reasons why we have inconsistent performance when we think we have processes. If we deal with this inconsistent performance by thinking we just have to make an adjustment to the process, we will never obtain our objective. Instead we need to treat the actions as a practice and work on practice solutions rather than trying to better specify one more aspect of a process.

Second, and more importantly with respect to our main interest in this book—PLM—we need to define what the implications are for using the approach. Unlike transaction-based systems that only deal with processes, PLM needs also to deal with practices. PLM applications will need to enable and facilitate practices, not simply control processes. PLM applications will need to deal with rich representations of information and present to PLM users the right information when they need it based on both search characteristics and contextual situations.

PLM will also encourage organizations to increase productivity and efficiency by examining the practices that operate within the organization and making tacit information explicit. This explicating and structuring of information will allow those practices to be moved to the right on the spectrum and made into processes.

There are many examples of the use of tacit information that results in a procedure being a practice rather than a process. One example is the use of free-form fields that really contain fairly structured information that two groups use to communicate with each other. For example, engineering always specifies to purchasing the vendor that they have worked with on requirements planning. Without making explicit the tacit information flow, the real problem comes when the systems group believes that this practice is really a process and fully automates it, directly linking the engineering system with the purchasing system. The practice linkage is lost and everyone scrambles to figure out why purchasing cannot procure the right products.

So, with respect to practices and processes, assessing and implementing PLM is a two-step process. Move practices to processes where possible by making tacit information explicit. This is especially important where the organization already thinks it has a process, but really has a practice. This will free people up to concentrate on practices, where context and judgment are important and where the procedure cannot be codified or made routine.

The second step is to ensure that PLM will enable and facilitate practices to make them more productive. The primary mechanism, as we will see in the rest of the book, is to be able to present the right information when it is needed—a very nontrivial task. If these two steps are successfully performed then, PLM will indeed be lean thinking and it will drive the next wave of productivity.

The Outline of the Book

- **Chapter 2:** What is PLM? How are the various thought leaders defining PLM? What models best describe PLM so as to capture PLM's most salient features?
- **Chapter 3:** What are the characteristics of PLM? Why are the characteristics important? What is the impact on organizations that these characteristics affect? How do these characteristics manifest themselves in PLM applications?
- **Chapter 4:** What is driving PLM? What are the trends in organizations that make PLM so compelling? What obstacles does PLM seek to minimize or remove? How should we evaluate and communicate the value of PLM from a financial perspective? How does this compare with other corporate initiatives, such as ERP.
- **Chapter 5:** What are the elements of PLM? What pieces have to be considered to implement PLM? What considerations of change within and outside of the organization need to be addressed?
- **Chapter 6:** It makes sense to start with PLM in design and engineering, but how? Where are the best places to start and what types of projects make sense? Do PLM projects make sense if developed only for a single functional area?
- **Chapter 7:** How are digital manufacturing and PLM connected? How do we break down the silos between design, engineering, and manufacturing? What areas of engineering should we focus on?
- **Chapter 8:** What are the opportunities for PLM outside the factory door? Why and how does PLM need to extend to the usage of the product? What implications does this have for design and quality?
- **Chapter 9:** Once we develop some experience, how do we leverage that into a PLM strategy? How do we balance the

short-term need to demonstrate the value of PLM with a longer-term view of the bigger opportunities that PLM potentially can deliver? How can we start carrying PLM projects over into other functional areas?

- **Chapter 10:** How do we assess an organization's readiness for PLM? What metrics do we need to collect and assess? What framework for assessing PLM readiness needs to be evaluated?
- **Chapter 11:** What are the real-world issues that PLM will need to confront? What issues, both technical and social, will need to be overcome for PLM to reach its full potential? What does the future hold for PLM? What technologies should we be concentrating on in order to fully enable PLM? What will be the implications of a PLM-enabled world?

Notes

1. The bits-for-atoms trade-off has been implicitly understood for a long time, although not necessarily well articulated. For an early view of this idea of living in a digital world, see N. Negroponte, *Being Digital* (1st ed.), New York: Knopf, 1995.
2. See P.S. Pande, R.P. Neuman, and R.R. Cavanagh, The Six Sigma Way: How GE, Motorola, and Other Top Companies Are Honing Their Performance, New York: McGraw-Hill, 2000.
3. This is in spite of controversial claims that IT has become a commodity. See N.G. Carr, *Does IT Matter?: Information Technology and the Corrosion of Competitive Advantage*, Boston: Harvard Business School Publishing, 2004. For a shorter version of the discussion, see the original article that generated a tremendous amount of controversy: N.G Carr, "IT doesn't matter." *Harvard Business Review*, *81*(5), 2003, 41-49. Also, see the letters to the editor in response to the original article by some very prominent academics and practitioners.
4. Personal conversation.
5. This is not to imply that this is the only substitution available. Machines, which are material, energy, and information, obviously can replace wasted time. (For a philosophical discussion of substitution sets see: A. Giddens, *The Constitution of Society: Introduction of the Theory of Structuration*, Berkeley: University of California Press, 1984.) The focus here is on the use of information as the key substitute. While it may be embodied in a machine, for our purposes, it does not have to be. In addition, the material and energy of machines are eventually "consumed," that is, used up and worn out. Information is not consumed.
6. In the case of information, it does not as of yet have any unit of measurement. While data is measured in bits or bytes, the information contained in the data is unmeasured, although one wag has suggested that it be measured in "wits."
7. In mathematical notation, we would represent the function for performing the costing as $C(x)$, where x is the amount of time (t), energy (e), or material

(*m*), used for a specific task. *C*(*x*) would operate by taking the *x* units and multiplying those units times the cost per unit, with the results being cost units (i.e., dollars, euros, etc.) Cost units are then directly comparable to other cost units. The cost function for information *C*(*i*) is not as direct, due to the lack of a unit of measurement for information. However, we can get the cost of information in the aggregate by using the costs of computers, software, instrumentation, and people's time to operate and maintain these systems as a proxy for the cost of information.

8. See C.E. Shannon and W. Weaver, *The Mathematica Theory of Communication*, Urbana: University of Illinois Press, 1949 for what is considered the seminal book on this topic.

9. See D.D Pattison and C.J. Teplitz, "Are Learning Curves Still Relevant?" *Management Accounting*, 70(8), 1989, 37-40. For Wright's original article see: Wright, T. P. Factors Affecting the Cost of Airplanes. *Journal of Aeronautical Sciences* 3(4), 1936, 122-128.

10. The economists' approach to experience curves has confused this issue. They often propose doubling production as a strategy for reducing costs. But doubling production does not cause reduced costs. It is the actions of the people involved in developing this information substitution that cause the reduction in costs. See W. W. Alberts, "The Experience Curve Doctrine Reconsidered," *Journal of Marketing Research*, 53(3), 1988, 36-49 for a discussion of this issue.

11. I qualify this statement because there are some extreme IT resource applications, such as weather forecasting, nuclear reaction simulation, crash test simulations. However, these applications are limited to a small group of users. PLM uses a substantial amount of IT resources for a wide community of users.

12. The Storage Law is a close corollary and storage capacity is thought to double every 12 months. See: www.creatingthe21stcentury.org/JSB2-pace-change.html for a concise description by John Seely Brown.

13. This is first-hand knowledge. As a teenager in the late 1960s, I was a systems programmer. In order to "cold boot" a GE 265 timesharing computer system that filled a 1,000 square foot room, I had to stand at the front of it and toggle in 50 or so 20-bit binary numbers. One missed bit and the system failed to come up.

14. P. Evans & T.S. Wurster, *Blown to Bits: How the New Economic of Information Transforms Strategy*. Boston: Harvard Business School Press, 1999.

15. A. Turing, "Computing Machinery and Intelligence," *Mind*, LIX(236), 1950, 433-460. No computer has of yet passed this test.

16. Even in moviemaking, the interest is in reducing art to process. Movie producers prefer "formulaic" movies over risky premises. Every successful movie spawns one or more sequels.

17. While "satisficing" is not in general use, it should be. It was coined by the Nobel-winning economist and computer scientist Herbert A. Simon. He used it to point out the myth of the economists' rational man who optimized his economic decision. Simon pointed out that with man's limited processing capability, he could not possibly optimize his decision making, except in the most limited situations. Instead, man satifices, by making acceptable decisions based on limited cognitive power. See H.A Simon, *The Sciences of the Artificial* (3rd ed.), Cambridge, Mass: MIT Press, 1996.

Constructing PLM

As IN ALL TECHNOLOGIES, PLM is a social construction with competing views and interests. It is difficult to capture all of the salient elements in a simple definition because there currently is a lack of shared understanding and agreement. This chapter will explore the various definitions of PLM to show the dispersion of views. The chapter will then propose some visual models and explain how they show rich explanations of what PLM is and can become. In addition, the chapter will explain PLM's relationship to the other major enterprise system, ERP.

New system and application concepts are not so much defined as they are constructed. Different constituencies, such as end users, software providers, academics, or research firms have different ways of looking at problems and solutions and try to tag them with new names that are reasonably descriptive of the concepts. The new name quickly becomes an acronym. However, unless there is resonance across constituencies, the effort dies out shortly or is adopted by a small subset of the potential group, with the concept having meaning only within a small band of experts.

However, if the acronym resonates with a broad enough group, then it starts appearing with more frequency. Companies with products to sell use the acronym in their trade literature and in their presentations. It is helpful to them because if these product providers can categorize their product as solving a class of problems that are implied by the acronym, then they can reduce their amount of sales effort. Instead of having to define both the solution space and the manner in which their product addresses it, they can concentrate on the latter. Companies that define their own problem space—for which they have the solution—often look as if they have products looking for problems that may not really exist.

The industry press also adopts and promotes the acronym. The term *industry press* covers the newspapers, periodicals, and magazines that focus on a particular discipline or specialty area, such as information technology, engineering, or manufacturing, or a particular industry, such as aerospace, automotive, or chemical processing.

This industry press is also predisposed to the creation of new acronyms. New acronyms are newsworthy and a reason for readers to seek information. The industry press views its role to be at the edge of new developments. It actively seeks out new concepts and technologies and helps its readers to assess their usefulness. One measurement of the adoption of new acronyms is the count of their appearance in the industry press. If the number of mentions does not show increases, it is unlikely that the acronym will find widespread adoption.

Companies with products that address the new problem space are also a source of new advertisers for the industry press. This raises the issue of bias. Most, but not all, of the industry press uses a free-to-qualified-subscribers model, in which their periodicals are distributed at no charge to subscribers who fill out qualification forms. Advertisers are the main source of revenue in this model. Although the industry press professes editorial independence and blatant touting would most likely result in loss of readership, the reality is that the opportunity for an advertiser to influence the reporting on a new acronym must be taken into consideration.

In addition to the industry press as an independent actor in the development of new acronyms and their underlying concepts, there are also the research firms. Research firms are organizations that are on a constant watch for new developments and assess their development and impact. They evaluate new concepts and technologies and produce reports and recommendations on these. These firms also evaluate products and their providers and produce reports describing and comparing these products and providers.[1]

The research firm's revenue model is different from the industry press revenue model. Research firms usually sell their research on a subscription or report basis. Unlike the industry press, research firms sell no advertising, so their focus is on the quality and usefulness of their reports. If the reports are inaccurate or promotional, then the buyers of those reports will fail to find them useful, and the research firms will lose clients. While the opportunity for bias is lessened because of the different revenue model, it is not altogether eliminated. Research firms also get some of their revenue from product providers who not only subscribe, but may also engage the research firm for a fee to evaluate products, research areas of opportunity, or speak at their conferences.

Another potential independent contributor to the development of new concepts and technologies is the academic institutions. In the Information Technology field, the academic institutions have been, at best, peripheral players in the construction process of new concepts, technologies, and their ensuing acronyms. While it is expected that there would be little bias in their perspectives, the academic community has had little participation in the process that results in the development of the new acronyms and their underlying concepts and technologies. While academics do develop research papers on these topics, most are written after the concept has been well established, and the practitioners driving the field rarely are acquainted with this research.

The final group in the construction process is the most potent. It is the users of the concepts and technologies. It is only if this group buys into the concepts and technologies and uses the acronyms themselves that the acronyms become established. If users do not devote resources to the area, buy products and services, subscribe to research reports, and read the industry press, then product providers very quickly change focus or go out of business; the industry press drops coverage; and research firms elect to cover other areas. To convince ourselves of this, we have only to look for "e-" anything in products, industry press, or research reports today.

A more successful and longer lasting concept that translated into technologies is Enterprise Resource Planning (ERP). Before the 1990s, there was no Enterprise Resource Planning. There may have been resource planning at the enterprise level, but before

1990, someone saying, "We have invested in an ERP system" would have brought a blank stare from the listener. Say that today, and a listener with reasonable business experience will have a fairly good idea of what the speaker is saying.

In the case of ERP, credit is given to a research firm, the Gartner Group, for introducing the acronym as part of a research report initiative.[2] As noted above, these new concepts and their associated acronyms do not appear out of thin air, but have their basis in other technologies and concepts. In this case, ERP is an extension, a substantial extension, to a manufacturing production concept called Material Requirements Planning (MRP), which was embodied in different versions of software. Gartner took the concepts of MRP and enhanced them so that they could apply to the whole production process throughout the enterprise. Although ERP and PLM are related, they have very different conceptual bases, as we will explore later in the chapter. Their acronyms came into existence through this process of consensus building by the constituencies involved in the area.

Defining PLM

References to PLM started to appear around the beginning of 2000.[3] While it may appear as though a brand-new concept suddenly appeared from nowhere, that was not the case. As with ERP, PLM emerged as a broader view of concepts and technologies that have been on the scene for some time. These concepts and technologies progressed along their own paths for a significant number of years. As they progressed and became more powerful and more mature, the opportunity to combine and reconfigure them in a new way that addressed a broader class of issues presented itself. These threads of other concepts and technologies were woven together to create PLM.

Among those threads of concepts and technologies, there are four that deserve the most attention as predecessors to PLM: Computer Aided Design (CAD), Engineering Data Management (EDM), Product Data Management (PDM), and Computer Integrated Manufacturing (CIM). Before we trace their contributions to PLM, let us look at where we are today in defining PLM.

When confronted with the question, "What is PLM?" the usual response is to try and reply with a definition. Fortunately, we have a few to pick from. The first one we will try is one that appeared in *CIO Magazine*:

> **Product lifecycle management is an integrated, information-driven approach to all aspects of a product's life, from its design through manufacture, deployment, and maintenance—culminating in the product's removal from service and final disposal. PLM software suites enable accessing, updating, manipulating, and reasoning about product information that is being produced in a fragmented and distributed environment. Another definition of PLM is the integration of business systems to manage a product's life cycle.[4]**

As we look at the definition, the most striking thing about it is the definitive manner in which PLM stakes out its territory. It is about *"all aspects of a product's life, from its design through manufacture, deployment and maintenance—culminating in the product's removal from service and final disposal."* One aspect of this phrase paints a broad picture: *"from its design through manufacture, deployment and maintenance—culminating in the product's removal from service and final disposal."* This is a bold claim. PLM is not simply an engineering, or a manufacturing, or a service initiative. The definition eschews a functional orientation for a cross-functional approach.

The definition, as presented here, is not even constrained to an organization, but implies that PLM manages information about the product no matter where the information resides. It is a rare organization today that designs, sources from itself, manufacturers, services, and disposes of a product within its four walls, no matter how vertically integrated it might be.

With new outsourcing models, product information can no longer be thought of as existing only within the boundaries of the organization, even for what have historically been treated as core functions, such as product design and development. In some cases, more individuals will deal with product information outside the organization that owns the product than will individuals inside that organization.

The immense scope implied by the definition of PLM is further reinforced by the amount of time that this definition covers.

The time period covered is from design to disposal, from cradle to the grave, or the picturesque from lust to dust.[5] This can be an extremely long period of time. For some durable goods, from design to disposal could be at least 50 years. In fact, 100 years or more is within the realm of possibility for building products such as elevators and industrial products such as grinding or milling machines. Many automobiles and trucks are in existence for this time frame, even if the majority are not used in everyday life. PLM is staking out a more substantial, cross-longitudinal inter-organizational view of information than has been done in the past.

Since information does not degrade, the duration of the information about a physical product is not an issue. However, the hardware and software systems that store and access the information are another matter. Hardware and software systems become obsolete and unsupported. Media degrades. PLM has signed up for a task of long duration. From an implementation standpoint, it is too early to determine whether PLM can meet the challenge and whether, as advances in hardware and software occur, the old information can migrate along with these advances.

While the longevity focus of PLM remains an open issue, another aspect of our PLM definition narrows the focus in another way. PLM is about the product. It is not about supply chains, employee knowledge management, or domain expertise. PLM is all about the product and its associated information. While these other things may be important and may tie into PLM, they are not directly the concern of PLM.

This means that a singular focus on PLM will not solve all the organization's issues. Organizations will need to plan and deal with issues regarding customer requirements and information, employee knowledge bases, domain expertise in the physical sciences, and supplier information. As will be discussed later in the chapter when we look at how these various areas fit with PLM, we will need to make sure that the systems that support these areas work well together.

The other aspect of the definition that deserves attention is the use of the term *information-driven approach* in answering the question, "What is PLM?" PLM is not just an application or even an entire software system. It is not only a supporting system for organizational practices. It is not just a strategy, nor is it just a philosophy about how we organize information. PLM encompasses all of these.

While *approach* is a bit nebulous and not very precise, it is serviceable enough to convey all the aspects of PLM that we have just described. The word we really want is *technology*. But to too many people, technology is synonymous with some tangible artifact, like a machine, electrical device, or some software code. If we use the word *technology* in our definition, the risk that people will think that they can simply go out and acquire a computer with software and have a PLM system is high enough to make the use of the word *technology* unacceptable.

But the definition stated above is not the only definition of PLM. Other organizations have proposed definitions that we need to examine and evaluate. The National Institute of Standards and Technology (NIST), the standards-setting research group of the United States Government, has published the following definition:

A strategic business approach for the effective management and use of corporate intellectual capital.[6]

While this is not an inaccurate definition, it is fairly vague and does not lend enough specificity to PLM. While product information is critical intellectual capital of a corporation, there are other kinds of corporate intellectual capital that are not related to PLM. Other kinds of intellectual capital include domain knowledge regarding customers, suppliers, and production processes, and research domains, such as materials properties, chemical reactions, etc. This definition also does not address the lifecycle aspect of PLM.

However, the NIST definition is consistent with the pervasive view that PLM is not only an application or a system. In addition, the NIST definition emphasizes the issue that product information is a substantial capital asset of the organization and that there must be attention to its capture for continued use by the organization.

PLM stands for Product Lifecycle Management, which is a blanket term for a group of software applications used by engineering, purchasing, marketing, manufacturing, R&D, and others that work on NPD&I.—AMR Research.[7]

As mentioned earlier, there is neither complete consensus nor closure regarding the definition of PLM. The particular definition quoted above reflects that lack of consensus and presents a different view of PLM. It represents PLM as software applications, which limits its definition substantially.

This definition also reflects one of the most prevalent alternate views of PLM. That view is that, while PLM is cross-functional (used by engineering, purchasing, marketing, manufacturing, R&D, and others), it is limited to new product development and introduction (NPD&I). While at odds with the much broader view of PLM that a number of the significant players within PLM have coalesced around, it has some merit for the following reasons.

1. The initial focus of PLM generally is on new product development. It is a fairly logical proposition for an organization to start a major new initiative at the beginning of the lifecycle of its new products. Some organizations might load existing items into PLM applications, but it will probably be more for the reason that these items can be used as components of new products than for the reason they can be used in other phases of the lifecycle, such as sales and service.

2. The state of the art in PLM applications has been focused on the new product development area. As seen below, some of the technologies that form the basis of these PLM offerings have been in development and use for many years. Software applications and systems for addressing later aspects of the products' lifecycle, such as manufacturing, sales and support, and disposal, are recent introductions and are relatively immature. While the prognosis is promising, it still remains to be seen whether these new additions can fulfill the promise that a wider definition of PLM proposes.

3. There are compelling business reasons for focusing on the NPD&I area. The estimates are that around 80 percent of the cost structure of a product is defined at the requirements definition and engineering stage of a new product's development. If so, it makes good business sense to focus on this area of the product's lifecycle when its lifetime cost structure can be impacted.[8]

A strategic business approach that applies a consistent set of business solutions in support of the collaborative creation, management, dissemination, and use of product definition information across the extended enterprise from concept to end of life—integrating people, processes, business systems, and information.—CIMdata

CIMdata, a research firm that has been focused on PLM and its predecessors for a number of years, has a definition that is similar to *CIO Magazine's* definition. CIMdata's definition is clear about the breadth covered by PLM. It is from concept to end of life. CIMdata stakes out the same broad territory for PLM and expects that PLM will be useful beyond the product introduction.

CIMdata also uses the vague word, *approach*, to capture all the various elements that go into PLM. However, CIMdata does clarify what the elements of such an *approach* might be. CIMdata refers to *integrating people, processes, business systems, and information.* This is a variation of a common theme regarding information systems. That theme is that information systems are made up of three distinct elements: people, processes, and technologies. CIMdata provides a variation on that theme by substituting business systems and information for technologies. It is difficult to make the argument that information exists as a separate element rather than as the result of people, processes, and business systems, but we will revisit these elements later in the book.

We will look at one last definition from our most important PLM constituency, the user. The Ford Motor Company was an early adopter of both PLM and its predecessor technologies. Ford crystallized this effort with the formation of an office called C3P in the mid 1990s. The three Cs in C3P refer to computer-aided design, computer-aided manufacturing, and computer-aided engineering (CAD/CAM/CAE), while the P refers to product information management (PIM), a term that was auditioned for what became PLM but failed to find the required support. Ford has been very influential with both the solution providers who develop the software for PLM and with other PLM users who look to Ford's experience in developing their own approach to PLM.

Ford's definition of PLM as presented at the University of Michigan's AUTO^e 2004 Conference for IT, engineering, and manufacturing automotive executives is

Product Lifecycle Management:

> Is concerned with processes, methods, and tools used from a product's inception through the end of its service life
>
> Is the science of bringing these three disciplines together to create an environment that enables creation, update, access, and, ultimately, deletion of product data
>
> Extends across traditional boundaries, PD–Manufacturing, Inter–Intra Enterprise, North America–Europe...
>
> Is not CAD, CAE, or any other discipline that exists to author or analyze narrow subsets of product data

Here again we have the consistent but bold claim that PLM deals with a product from its inception until its end of life. The definition further reinforces this idea by affirming that it is not Computer Aided Design (CAD), Computer Aided Engineering (CAE) or any other subset of product data. As stated above, while this is an ambitious and bold statement, the participants in developing the vision of PLM are fairly consistent that PLM needs to deal with product information for the entire life of the product. This is without firm assurance that this can really be accomplished at this stage.

The Ford definition refers to a triumvirate of processes, methods, and tools. This is similar to the people, process, and technology theme, with tools and technology being fairly interchangeable. Ford's use of methods in place of people is interesting. It also may better describe the key elements. Simply having people is not enough to create information systems. It has to be people doing something. The methods or, as we used in the first chapter, the practices that people employ in working together are what generates this information. Processes are not enough. We need methods or practices.

The Ford definition also emphasizes the spanning of boundaries. It mentions two: functional boundaries and geographical boundaries. We will see that this is a major emphasis for PLM: the attempt to counteract the creation of information silos brought upon by the division of labor by using PLM to coordinate functional and even geographic areas.

One last interesting observation is that the Ford definition calls for the deletion of product information. This is not seen in any other definition of PLM. It is an interesting idea.

Finally, what can we conclude about the purpose that underlies all these definitions of PLM? It is that by creating and using this product information we can make more efficient use of our physical resources. We can trade this information for wasted time, energy, and material. Using this product information across the organization is driving the next generation of lean thinking. Lean thinking is not confined to manufacturing, but expands across the entire organization and into the supply chain.

So what can we conclude about defining PLM? Here are the main elements that we can say need to be included in PLM:

- PLM is about product data, information, and knowledge.
- PLM concerns itself with the entire life of the product, from inception to end-of-life.
- PLM is an approach that is more than software or processes.
- PLM crosses boundaries: functional, geographical, and organizational.
- PLM combines the elements of people in action (practices or methods), processes, and technology.
- PLM drives the next generation of lean thinking.

The definition of PLM we will use through the rest of the book is:

Product Lifecycle Management (PLM) is an integrated, information-driven approach comprised of people, processes/practices, and technology to all aspects of a product's life, from its design through manufacture, deployment and maintenance—culminating in the product's removal from service and final disposal. By trading product information for wasted time, energy, and material across the entire organization and into the supply chain, PLM drives the next generation of lean thinking.

PLM Lifecycle Model

The problem with definitions is that they presuppose a fair amount of knowledge about the topic. The definition of a car from *The American Heritage® Dictionary* is "A self-propelled passenger vehicle that usually has four wheels and an internal-combustion engine, used for land transport. Also called motorcar."[9] The definition of a car from *Merriam-Webster* is "a usually four-wheeled automotive vehicle designed for passenger transportation."[10] These definitions make perfect sense to us. We immediately picture the four wheels in the rectangular configuration, the driver sitting on a seat in the front-part of the car. In other words, we know what an automobile's normal configurations are, and the definition evokes it as much as describes it.

However, consider the alien archeologist of the far future who finds only our dictionary. Who knows what configurations that he/she/it might come up with as it tries to understand what an automobile is from simply parsing the definition. We are not unlike that alien at this stage of PLM, as we attempt to define it. Because PLM is in its early stages, there is no common view of PLM for our definitions to evoke.

Add to this that, as evidenced by the definitions presented above, there are differences in definitions from experts working in the area. Therefore, we should not be surprised that attempting to explain PLM to executives and decision makers by way of definitions would be fraught with difficulty. Even if we simply selected one definition, the additional descriptions that we need to develop to describe PLM sufficiently are voluminous. Our conundrum is how to convey to these executives and decision makers a succinct but complete description of PLM.

A potential answer to this lies in the old adage, "A picture is worth a thousand words." To help in the understanding of PLM, we have developed visual models. Visual models help us understand concepts because they provide a rich amount of information at a glance. We can see a variety of relationships laid out, and we can make assumptions about how the various pieces fit together. While some models are extremely detailed, the more powerful models for understanding broad concepts are the simple ones.

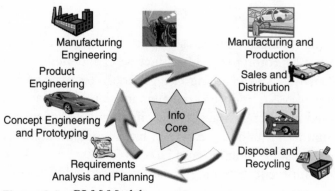

Figure 2.1   PLM Model

Figure 2.1 is our representation of the PLM Model. At the center of the model is an information core. This core represents all the product data and information about the product throughout the product's life. The information core is separate from the functions or stages that use it. The product information does not belong to any one functional area, but is available to all functional areas.

Around the informational core are the functional areas that comprise a product's lifecycle. These functional areas are how organizations divide up the major categories of a product's life: plan, design, build, support, and dispose.

Plan

The model starts with requirement analysis and planning, which is the initial step in developing any product. What are the functions the product must perform? What are the requirements that the product must meet? These requirements are mapped into specifications. A requirement for a power window maps to a specification that a sheet of glass of x pounds must be raised and lowered y inches in z time for q cycles. In some cases, these requirements are obtained directly from the customer who is buying the product. In other cases, the requirements are arrived at indirectly through a marketing or entrepreneurial function that makes the determination that a customer will buy a product that has certain functions.

Design

The requirements mentioned above are taken up by concept engineering and prototyping. The functions generated by requirements analysis and planning can generally be realized in more than one way. In addition, the functional requirements can be combined in different ways so that the number of parts required for a number of functions can vary. Each function might have its own part, or some functions can be combined into a single part. In addition, different technologies can be employed to produce the same functionality. For example, electronic controls may replace hydraulic ones.

The aesthetics of a product come into play at this stage. While "form follows function" is a basic tenet of product design, the function might allow for a wide variety of forms. Automobile designers and consumer product designers spend a great deal of their product design effort and produce a wide variety of final design forms for the same functions. However, as these designers work, they need to make sure that their concepts and prototypes meet all the mandatory functional requirements and as many of the desired functional requirements as possible. When trade-offs arise, as they invariably do, the concept and prototype designers must share that information with the specification developers.

The next step in the cycle, product engineering, takes these functional designs and prototypes and turns them into exact specifications. It is product engineering's role to fully specify the product at this stage. While the concept engineers developed the general form, the product engineers take that general form and completely specify it. The product engineers must make sure that all the various components fit together in an integrated system and that the system is internally consistent. In addition, the product engineers must run various tests, such as force and flow analysis, on components and the entire product to insure that the product really meets the requirements specified. The product engineers need to add this body of information to the product lifecycle model. At the end of this phase, the components that make up the product are fully defined in a math-based model or CAD specification.

Build

Once the product is fully specified, it is the role of manufacturing engineering to determine how the product must be built. The designs must be analyzed and the bill of process (BofP) developed to specify what operations must be done in what sequence to create the desired part. Those parts must then be assembled in a specific sequence to develop the completed product.

In most organizations, this entails three fairly distinct phases: building the first product, ramping production up, and building the rest of the products. In manufacturing the first product, there are two considerations: building it in a new plant or building it in existing plants.

If the product is to be built in a new plant, then the design phase just got substantially more complicated, because not only must the product be designed, but the tools to be used in building that product must be specified. Manufacturing engineers must either select tools that will meet these specifications or send out Requests for Quotations (RFQs) to tool manufacturers to ask them to provide quotes on building tools that will meet those specifications.

If the product is to be built in an existing plant or plants, the task does not get much easier. In what is akin to a linear programming problem, the manufacturing engineer must determine how to manufacture the parts given the tools and equipment existing in those plants. In some cases, this will necessitate changing the design and/or the Bill of Process if one design cannot be built with the existing equipment but an equivalent design can be. In addition, if multiple plants exist with different equipment configurations or Bills of Systems, then the task of the engineer is to determine which Bill of Systems matches up best with parts and products to be built.

During this process, there may be issues that affect the design. One that was mentioned above is the equivalent design issue, where one design can be built while the other cannot. The other issue is that the design cannot be built as it is designed. In some cases, the issue is so severe that it must be sent back to the design engineers because tooling conflicts cannot be resolved.

Other issues are handled at the manufacturing level. A common one is to find holes cut in a surface that is shown to be uncut on the design specifications. The reason is that the manufacturing team needed access to install a bolt or fastener. While this may sound expedient, cutting this hole may change the characteristics of the strength of that surface. This is something that is never subject to analysis because the design engineers do not know it exists.

The manufacturing and production stage is when the rest of the products are built. As explored in the first chapter, experience curves suggest that a good deal of experimentation goes on during this process.[11] The manufacturing and production staff takes the product plans and the Bills of Process and continually revises and refines them to build the parts and products using the least resources.

Support

The sales and distribution function uses the product information (1) to tell the buyer and the user of the product what the functions and specifications of the product are and (2) to keep the product performing to those expected specifications. The user of the product needs product information to understand how to obtain the required function (e.g., push this button to adjust the seat this way). The service organization that is either associated with the product producer or independent of it needs product information to service the product in case of malfunction or failure.

This part of the product lifecycle is also a potentially rich source of information about the product. How the product actually performs in use is the best information with which to determine if the product is designed properly. Warranty data and, even better, data collected by monitoring systems are important information about whether a product performs its functions as specified.

Dispose

The final aspect of the product's lifecycle, disposal and recycling, close out the product's life. Information about how the product was designed and its component makeup is necessary for effective and efficient recycling and disposal. Information about whether the product could be recycled using the processes designed when it was built is important information for future product designs.

The cycle starts all over again with the next version of the product, building on that information core. While the model shows this sequential process that reflects how we see the product lifecycle, reality is a bit messier. The engineering and manufacturing functions often need to iterate designs to a solution that meets the functional requirements and can actually be built at the right cost. Usage experience in the field can point to a major problem that needs to be fixed not only in the new products, but also in the existing ones.

However, in spite of these realities, the PLM Lifecycle model is a good one. It reflects the fact that there are various separate stages to a product's life. It reflects that these stages are relatively sequential, and that this core of information about the product is essential for the effective development and use of the product through time. Product information needs to be developed and used throughout the entire life of the product and not compartmentalized by stages or functions. This visual model is a great deal more descriptive than our short definitions.

The Threads of PLM

As stated above, the perception is that PLM appeared fully formed within the past few years. However, it is unlikely that a concept that requires complex technologies to enable it formed virtually overnight. What is more likely and supportable is that PLM took existing threads of concepts and technologies that existed in their own right and wove them together to form this new concept.

If we take a close look at PLM, we can discern some of the conceptual and technological threads that have gone into forming PLM. Much like a fabric woven from threads, PLM is more than the sum of these threads, and it is how they are woven together that makes PLM such a powerful enabler for the next wave of productivity. The threads we will examine are: Computer Aided Design (CAD), Engineering Data Management (EDM), Product Data Management (PDM), and Computer Integrated Manufacturing (CIM).

Computer Aided Design (CAD)

Computer Aided Design (CAD) refers to the math-based descriptions of the physical representation of products. Math-based descriptions use a system of mathematical descriptions to locate and consistently replicate shapes in either two-dimensional or three-dimensional space. These systems started out as simple drawing systems to aid the designer in producing faster and more accurate drawings, hence, the name computer aided design. However, CAD systems rapidly developed functionality that moved them from a periphery system augmenting design to the focal system enabling and controlling design.

Using geometric figures, CAD representations attempt to provide a complete definition of the physical structure of a product. Because these are complete specifications, they are available for the same types of measurements as an actual physical object. In addition, internal measurements can be computed as easily as external measurements.

In fact, these CAD representations are better than an actual physical object because, being math-based, the CAD specifications do not deviate from implementation to implementation. Specifications can be developed with as much precision as is designed into the CAD software. Each replication of the math-based representation will be exactly the same each and every time. Measurements will also be absolutely consistent. The distance between any two points can and will be measured to exactly the same decimal point each and every time, no matter which computer system it is represented on.[12]

CAD applications are major feats of software engineering. CAD systems take the rules of physical structures, lines, vectors, surfaces, etc. and attempt to mimic them within the memory space of a computer system. As the richest description of product information, CAD representations are at the heart of PLM.

Once the purview of specialized hardware/software systems, CAD representations began as two-dimensional (2D) representations or as views of product information. CAD systems were only worked on by specialized designers to provide the output, usually two-dimensional paper drawings that engineers would use to guide their decisions about product structure. As described in its name, CAD representations initially were only an aid or assist to the designers and engineers.

At some point in time, the process changed. Rather than have designers handing rough drawings and specifications to CAD specialists for input and refinement in a CAD system, the designers themselves sat down at the computer console and designed their parts and products directly on the CAD system, experimenting and refining designs as they worked. This marked the transition from the system's being an "aid" to design to the system's being an authoring tool, although the CAD name persisted.[13]

The output of these CAD systems generally had to be realized in a prototype or mock-up—a three-dimensional (3D) physical representation of the CAD specifications. Except for the simplest of designs, the parts represented by the drawing had to be physically constructed so they could be examined for fit and interference with other parts, for consistency across the different two-dimensional views, and for manufacturability by the tools on the factory floor.

The value in these early CAD systems was as a producer of consistent and quality drawings, where changes could be built upon the existing specifications that were saved in the CAD system, rather than starting all over with a new drawing each time there was a change to the product specifications.[14] However, the mental work in translating these designs into 3D representations took place in the minds of the designers and engineers and, until it was translated into a three-dimensional object, it was difficult for multiple designers—unless highly experienced and talented—to have the same mental representations.

But, over the years, as the software developed more capability and standard hardware advanced to the point where the specialized hardware was no longer necessary,[15] these CAD systems advanced to where they were not simply for designing and printing consistent drawings. The CAD representations could be distributed more widely in the organization so that engineers could start pulling data from the representations for their own purposes.

The impact of the ability of CAD systems to develop three-dimensional representations cannot be underestimated. While it seems to be an evolutionary development—two dimensions plus one dimension make three dimensions—it is a revolutionary conceptual advance. The mental imagery required by designers and engineers to "see" three-dimensional objects from two-dimensional renderings can now be done by the computer.

And done better. What two-dimensional CAD implementations required was that mock-ups, clay-renderings, or similar three-dimensional models actually had to be built in order for the designer or engineer to "see" the design for all but the most simple objects. In addition, taking measurements to validate the design had to be done in multi-step fashion as the measurements moved from one geometric plane to another. Two-dimensional measurements could be done fairly easily. Three-dimensional measurements were much more difficult and prone to errors in translation from two dimensions to three dimensions.

Once designs could be rendered in three dimensions, it now became possible not only to produce two-dimensional drawings, but to generate an image of what the actual product would look like. CAD systems can now create three-dimensional representations of physical objects, rotate them in space, match them up, detect where they physically interfere with one another, and calculate their dimensions with the most significant precision possible.

This makes "virtuality" a possibility. While there is much more to virtuality, as we have defined it under the Grieves Test, three-dimensional representations are a fundamental condition. In order to make a representation of an object that is indistinguishable from its tangible counterpart, it needs to be able to be rendered in the same dimensions. Two dimensions just do not suffice.

In addition, the breakthrough to three dimensions made possible Computer Aided Engineering (CAE).[16] Data could be extracted from the math-based descriptions to perform other functions, such as Finite Element Analysis (FEA), which tests for structural integrity and performance. Visualization programs were developed that could allow individuals other than the designer running special CAD software to see, spatially manipulate the designs, and even do limited modifications of the designs to perform fit and function analysis with related designs.

Once these math-based representations are developed, their representations are only limited by the technology available to present them. Thin representations on paper, while still in use, are becoming a thing of the past. Large-screen displays provide dynamic representations of the math-based products. General Motors, as well as most automotive manufacturers, has a "power wall" in its design studio that allows a life-size representation of a car to be displayed. Design and engineering management can examine in gross or minute detail the actual car, which exists only as an aggregation of math-based descriptions.

While private networks have connected CAD systems for many years, the availability of the Internet over the past decade has made the sharing of the math-based product descriptions available on a much wider basis, not only within an organization but throughout the supply chain. As available bandwidths have continued to increase, the using and sharing of this data have accelerated.

Engineering Data Management (EDM)

While CAD specifications are the heart of PLM, there are other key data elements and information about the product that is equally important. Math-based specifications describe the products from a geometric perspective, but do not fully describe them. The math-based information needs to be augmented by other information, called *characteristics*. Characteristics can be described as any information that is needed to describe the product and include tolerances, tensile strength, weight restrictions, adhesives, and conductivity requirements. These characteristics must be associated with the math-based or geometric information to fully describe the product.

Also, there is other information in addition to the information that describes the product itself, which needs to be associated with the product. Information such as the process for building the product, the methodology for coating or painting the product, the methodology for testing the product, the instruments and processes required to carry out the testing methodology, and the results of the testing procedure needed to be associated and managed with the product and was the focus of Engineering Data Management (EDM).

Engineering Data Management developed from a direction that was opposite to that of CAD systems, which were specialized, purposely built, integrated applications. While there were a number of specialized programs for EDM data collection, tracking, and reporting, the program predominately used by engineers was and is still a standard and common application.

However, that standard application is Microsoft's spreadsheet, Excel, which allows for data organization in an infinite variety of formats. So, while engineers maintain their engineering data in the same application, the probability that it will be in the same format is extremely remote.

Each engineer or group of engineers developed and kept their engineering data in different forms, often in different forms for different projects and programs. While there have been some efforts to standardize the formats within groups, divisions, and even organizations, the reality is that even the standard formats have been customized by the individual engineers using these formats.

Although the personal productivity of the engineers increased through the use of these tools, the next wave of productivity—organizational productivity—remains hampered by the inconsistencies of the various personal implementations. The largest database of engineering data is in a common database—Excel. But it remains relatively unavailable for organization-wide usage. So, while engineers had extensive information on cost data, product specifications, and engineering changes, this information needed reorganization to be useable on an organizational scope.

In terms of function, EDM consists of the data and information that abstract, define, and describe the product. While structure of the product is obviously implicit in the math-based descriptions, it is much too cumbersome, inefficient, and computer-intensive to inspect the math-based product description any time there is an issue of product composition to deal with.

The answer to this was to abstract and identify discrete elements that make up the product composition. These identified and named elements then stand in as a reference for the math-based representations. These elements can be ordered by their relationship to other elements and diagrammed to show relationships.

This is, of course, the Bill of Materials (BOM), which shows the same structure, composition, and relationship as the math-based representations. But it does so in an abstracted manner that human beings, with limited processing capability, can handle. Figure 2.2 shows a simple bill of materials for a fan. The lowest level is an item, part, or component. All three descriptions will be used interchangeably throughout the book unless otherwise qualified. These items, parts, or components are organized into assemblies. Assemblies are logical groups that perform specific functions and can usually be manufactured, assembled, and tested as a unit. Assemblies are combined to form products.

The BOM is an excellent structure to stand in for the actual description of the product. Additional information, such as characteristics, can be and needs to be associated with the parts that make up the product and is not derivable from the math-based descriptions. This information includes cost, weight, source, material composition, and availability. By associating this information with the named parts, necessary calculations can easily be done. Answers to questions such as, "What is the cost of this assembly of parts?" or "What is the weight of this product?" can readily be computed.

An issue with abstracting the structure information from its math-based representation is that there can be more than one BOM structure that can map to the product structure and still be correct. The classic example is that of Figure 2.3, which shows a part that has a groove down the middle. One way to make this part is to take a solid block and cut a groove. Another way is to take a base and add two rails, leaving the center open. Two BOMs, one with a single piece specifying a milling operation and one with three pieces specifying an assembly operation, meet the requirements of the product description. While in this case the weight

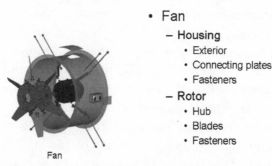

- Fan
  - Housing
    - Exterior
    - Connecting plates
    - Fasteners
  - Rotor
    - Hub
    - Blades
    - Fasteners

Fan

Figure 2.2   Bill of Materials

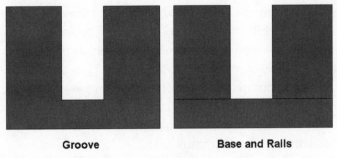

Groove   Base and Rails

Figure 2.3 One Design, Two Implementations

might be the same, the costs of not only the materials but the capital equipment necessary (a milling machine in one case but not the other) could differ significantly.

If this simple example has two conforming BOMs, we might expect more complicated products to generate many more alternatives. This is indeed the case. The problem is particularly prevalent across functional areas. In the example above, engineering might specify the part with a groove, but manufacturing makes it with a base and two rails. In most organizations, there may be at least three conforming but incompatible BOMs, one for engineering (EBOM), one for manufacturing (MBOM), and one for the accounting/finance department (FinBOM).

Product Data Management (PDM)

PDM applications developed as a way to organize CAD and EDM disparate formats in databases that were prestructured. PDM systems arose as a way to organize and catalog the proliferation of CAD files that an organization began to accumulate rapidly. It was no longer feasible or even useful to view the production of the CAD system, the actual drawing, as the end product. The CAD designs themselves were the intellectual property, and rendering them on paper only became one of a number of possibilities.

However, for the most part, these PDM applications were simply repositories of information that required the user to develop manual processes to make use of the information contained in them. While some of these systems began to incorporate some processes in the form of workflows, the majority of the systems were, as the name implies, focused on only managing the product data.

A primary purpose of PDM systems was to replace an organization's reliance on paper and microfiche archives with a reliance on electronic archives. Up until the development and deployment of PDM systems, most organizations had repositories that consisted of rooms of archived paper or microfiche drawings. The manual storing, accessing, and maintaining of these paper archives were inefficient and costly, even if the requirement for this information was entirely local. In addition, once the product information moved into the electronic realm, the discipline of filing a paper version with the centralized repository declined in some organizations. The thought was that the electronic version existed on the CAD station or on some backup.

Once we added the requirement of geographically dispersed groups, internal or external, needing access to this information, the costs skyrocketed while the efficiency plummeted. PDM introduced the idea that the product data could safely and effectively be organized, maintained, and accessed in native digital form.

In addition, the focus was on managing the product data only in the design and engineering function. The product data that PDM concerned itself with was only of use in the engineering function. There was little or no focus on using this product data in other functional areas. Another limitation was that PDM was targeted at a single organization and sometimes only for subsets of that single organization. The use of this product information was not intended for the wider supply chain.

Some contend that PLM is simply renamed PDM.[17] This may seem to be the case to those who are only looking at engineering functions. However, as we will see in the remainder of the book, PLM is qualitatively a different concept than PDM. It was, however, these PDM systems that took the first step in proving the feasibility and usefulness of aggregating different types of product information in digital form. PDM was an instrumental first step and formed the basis of the PLM systems that have superseded PDM applications.

Computer Integrated Manufacturing (CIM)

Computer Integrated Manufacturing (CIM) has an extensive history and was recognized early on for the promise that information could and should be shared across functional areas. Specifically, data and information from engineering could be transferred to and used by the manufacturing function in an electronic format. CIM embodies the idea that a computer system could integrate the functions necessary to design, engineer, and manufacture a product.

This extended the idea of Computer Aided Manufacturing (CAM), which had a much simpler premise of using math-based CAD descriptions to generate numerical control (N/C) programs (which are the programs that control the actions of automated manufacturing machines). In its advanced form, CAM data would also drive machine sequences and product routings. However, CAM implied a perspective for a single product, whereas CIM implied a broader view of all an organization's products against all the organization's production facilities and resources.

Both of these concepts were intended to breach the traditional barrier between engineering and manufacturing. The old method of the design and engineering department's developing the product and throwing it over the transom to manufacturing, which then had to determine how to build the product, was grossly inefficient. Manufacturing consistently had to re-create information that was readily available in the design and manufacturing phase, but was lost in the hand-off. In addition, design and engineering often developed designs that were not buildable. CIM applications hoped to take design and engineering data and directly drive the machines and production processes on the factory floor.

The idea behind CIM that cross-functional use of information can reduce duplication of effort and waste and drive productivity is a major aspect of PLM. However, CIM is acknowledged by most students of the area as never having lived up to its promise. The computing technologies available at the time were not up to the task, so CIM was limited in its application. In addition, CIM in some ways was broader than PLM, because it also encompassed enterprise resource planning (ERP) and supply chain management (SCM) functions.

While CIM is focused on manufacturing, other functions such as after-market service and disposal can also benefit from the sharing of data and information across functional areas. PLM was envisioned by many to be the common concept that pulled together math-based data, engineering data, and other information about the product and used it across other functional, nonengineering areas, starting with manufacturing. In addition, PLM has limited its scope to product information and works with—not in place of—ERP and SCM.

PLM starts with these predecessor technologies as major threads and some other concepts and technologies (like collaborative engineering and portfolio project management) as minor threads. PLM then attempts to weave a composite and cohesive view of product information.

Weaving the Threads into PLM

What enabled the weaving together of these threads of different but complementary technologies and concepts into PLM? The answer is the march up the trajectory of computer technology development. Continual increases in computing power, storage capacity, and communication bandwidth enabled these technologies to keep increasing in scope and functionality so that they were able to advance in capability. This capability could be combined into a more powerful technology. Increases in computing and storage capability allowed CAD designs to move from being the specialized, expensive systems of designers to the desktops of any designer or engineer who either needed or simply wanted to examine the product specification.

In the early CAD systems, when communications throughput was limited, it was only feasible to print out the CAD drawings and physically distribute those drawings to the people who needed them. The ability to send drawings from one CAD system to another required expensive communication bandwidths. Sending the CAD information to non-CAD systems was all but impossible. The processing time to propagate a design change throughout a complicated product initially could take days. As the computing technology advanced, this dropped to hours and then became minutes.

When it took days, design changes were batched up. Designers in one area did not know the changes that affected their work in another area. After the batch of changes was processed, the time-consuming and often redundant process of reconciling inconsistencies had to be undertaken. With the computing power of today, updates can take place in real time, and everyone knows that he or she is working with the latest version.

While the authoring of the product design was the initial step in the development of a pipeline that would produce a stream of these products, it needed to be combined with other, nongeometric information about the product. This combination of geometric and nongeometric information, the informational core, needed to be made available to the other functions that would engineer, manufacture, support, and eventually dispose of the resulting physical product. The different threads of technology that focused on different aspects of this information needed to be woven together in order to produce this informational core that could be used by all the functions that needed product information.

What was impossible to do 20 years ago and what we struggled to do stand-alone a decade ago became a standard point solution a few years ago. That capability is now commonplace and can now become a component in PLM's larger perspective. It is these advances in computer technology that allow us to combine our efforts in technologies such as CAD, EDM, PDM, and CIM and develop the new, integrated approach now called PLM. It is these advances and our ability to capitalize on them that drive our next wave of productivity.

Comparing PLM to ERP

Thus far, we have attempted to develop a view of PLM that approached it from a number of different perspectives: a definitional view, a visual model view, and a composite view of predecessor approaches. When dealing with concepts with such broad scale and scope, it is often difficult to draw clear boundaries that differentiate conceptual areas that are so broad. With respect to PLM, a natural and common comparison is with another very broad approach to enterprise information, Enterprise Resource Planning (ERP).

Even if we are successful in defining what PLM is, there are inevitable questions that compare and contrast PLM and ERP. If we have an ERP system, do we need PLM? Are PLM and ERP competitive approaches? How do PLM and ERP work together? In this section, we will explore and answer those questions.

To do that we first need to look at how our organizations are structured. Most, if not all, of today's modern organizations are divided into functional areas. The common functional areas are design and development, engineering, production, sales and service, and disposal and recycling. In addition to being divided into these functional areas, organizations all possess domains of knowledge. Domains of knowledge are areas of distinct knowledge about certain things that have a common theme.

The most common domains of knowledge in an organization are knowledge about products, knowledge about customers, knowledge about employees, and knowledge about suppliers. Knowledge about products deals with all the information an organization has about a product: how it needs to be designed, how it needs to be manufactured, the functionality it needs to have, etc. Knowledge about customers deals with customer-specific knowledge: their requirements, their procedures for doing business, their ways of making decisions. Knowledge about employees deals with employees' knowledge of areas of expertise: the processes they perform, their expertise in certain areas. Finally, knowledge about suppliers deals with expertise of suppliers: products that suppliers have to offer, their manner of doing business, their quality of work, their reliability, and other such things.

If we lay out the functional areas against the domains of knowledge, as we've done in Figure 2.4, we see that we basically have a matrix with functions on the left-hand side and domains of knowledge running along the bottom. Product Lifecycle Management by definition encompasses the product domain of knowledge, and it crosses all the functional areas, as shown by the vertical bar. Product Lifecycle Management consists of all information that deals with the product in the design and development phase, the engineering phase, the production phase, the sales and service phase, and the disposal and recycling phase. PLM clearly is intended to match up with the domain of knowledge about the product and to encompass all the functional areas of an organization.

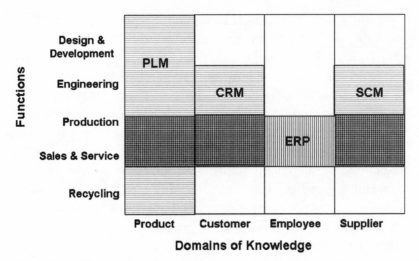

Figure 2.4 Comparing Systems

ERP, on the other hand, does the opposite. As shown by the horizontal bar, ERP crosses the various domains of knowledge—product, customer, employee, and supplier—but only focuses on the functional areas of production and sales and service. ERP is also primarily transaction-based. It is concerned with taking the information about a transaction, with respect to the product, the customer, the employee, and the supplier, and tracking that information in order to produce its unit of interest, a completed order.

As such, ERP is more narrowly focused, but concerns itself with more domains of knowledge in the organization. It is concerned with an end result, producing a transaction with respect to the delivery of a product or service to a customer. As we can see, however, ERP and PLM have an area of overlap in the areas of production and sales and service with respect to the information that they are concerned with. Because ERP is horizontal in nature, it also overlaps with supply chain management (SCM) and customer relations management (CRM). But again, these domains of knowledge are much deeper for the CRM and SCM applications and systems than the ERP system has to deal with.

Another way to get a sense of the commonality and differences between PLM and ERP is to look at the granularity of information that they deal with. In Figure 2.5, we have a matrix that increases the granularity of information that an organization deals with. At the most granular level is a transaction. Systems that are transaction based are only concerned with the specific transaction at hand. A transaction begins and ends as an event does, and then it is done with in the system.[18] Granularity increases, such that transactions can be part of an order. There may be a number of different transactions that are processed and shipped with respect to a single order.

Programs are the next step up in granularity. Programs are an entire series of product shipments that deal with a specific type of product, usually referred to as a model, or a model series if there are variations to the model. So, for example, there may be a program with respect to a car model that covers one or more years. The program may be focused on a jet fighter like Lockheed's Joint Strike Fighter (JSF) program that covers not only one jet fighter model, but variations on that jet fighter plane for the Air Force and the Navy.

Increasing granularity means that we move to a family of products, and these are products that are similar in definition, but differ in many respects. A family of trucks may have not only a number of different models within the truck line, but they may be built on a common platform that has substantially different names but is in the same family. Finally, there is the industry, which basically is the continuation of the family with a common technology into the future, with continuity.

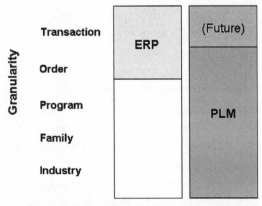

Figure 2.5 PLM and ERP in Granularity

As we can see, ERP is primarily concerned with the transaction and the order. Once an order is closed out, the ERP system processes the transactions with respect to that order, but is not very much concerned with the order beyond that. On the other hand, PLM's granularity is concerned with the order for the product and extends not only into the program, but into the family and the entire industry. In addition, we believe that the future definition of PLM will extend to individual transactions, referred to as *as-built*. Within PLM, each product will be serialized or quasi-serialized, such that the tracking of that individual product will continue throughout its entire life and reflect changes to the product as it continues throughout its lifecycle.

Using these matrices, we can see that PLM and ERP are complementary systems, and that by definition, PLM is more concerned with the domain of product knowledge and less concerned with the transaction than ERP is. This is not to say that ERP systems cannot incorporate PLM-like informational bases within them. In some cases, vendors of ERP systems would like to do that in order to be able to branch out their ERP systems. But what this does say is that on a conceptual basis, PLM and ERP systems are defined as complementary, although their implementations might attempt to encroach on each other's functions. It would seem to make sense, if we let each system have its own area of expertise, that all product information that ERP systems deal with would be in the PLM system. Given our interest in having only a singular view of the information as we will discuss in the next chapter, the ERP system ought not to duplicate the information that is in the PLM system. If we look at Figure 2.6, we can see a real-life example in which this is exactly how it works.

Figure 2.6 is an example of the Jet Propulsion Laboratory's (JPL) combination Enterprise Resource Planning system and Product Lifecycle Management system. On the right side of the diagram, the ERP system deals with the production of the product on a transaction basis. It controls the resources needed to produce a product through its processes. However, any information about the product is contained in the PLM system, as evidenced by the connecting lines between the ERP and PLM systems. In point of fact, no product information can exist in the ERP system. Any

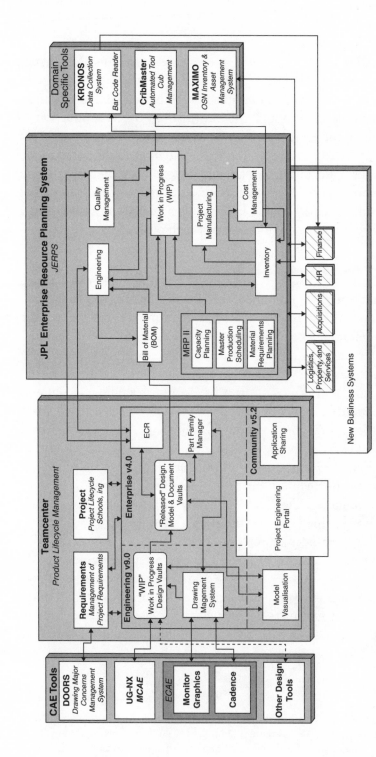

Figure 2.6 PLM/ERP Example. *Source:* JPL, Pasadena, CA.

information about a product has to be created in the PLM system before it can be referenced in the ERP system.

So we can see by this example that PLM and ERP are complementary systems, although the information for the product ought to reside in the Product Lifecycle Management system and not in the ERP system. However, the ERP system needs not only to reference that data, but also to use it at various stages in processing its transactions in order to do so with an element of efficiency. The end result is that PLM and ERP systems need to be aware of each other and work in conjunction with each other.

Summary

In this chapter, we examined how PLM is being constructed as a new way to think about product information. We surveyed the different definitions being proposed to describe PLM. We then proposed a richer way to describe PLM, the visual PLM Model, with the functions necessary to designing, engineering, manufacturing, supporting, and disposing of a product surrounding a common, informational core. We next showed how PLM has evolved out of and is the culmination of previous efforts in CAD/CAM, EDM, PDM, and CIM. Finally, we examined a commonly asked question: Do PLM and ERP perform the same functions? We concluded that they are complementary and not competitive approaches.

We should finish with noting that by constructing a view of PLM we are not finishing the task, but simply starting one. While we have woven together older approaches or technologies to form a new view of product information in PLM, PLM is an extensive framework. We can expect to see and are already seeing new initiatives coalescing out of the PLM framework. Collaborative Product Design (cPD) and Digital Manufacturing (DM), which we will discuss in later chapters, focus their attention back on a specific functional area, but do so using the framework and philosophies of PLM.

Notes

1. Examples of research firms with an interest in the PLM area are: CIMdata, AMR Research, ARC Advisory, D.H. Brown, and The Gartner Group. Their involvement with the PLM areas varies greatly by organization. CIMdata is

exclusively involved with PLM. The Gartner Group has many areas in the information technology space it follows, of which PLM is one.

2. See M. Endry and T. White, "What Is an Enterprise Resource Planning (ERP) System?" In D.R. Laube and R.F. Zammuto (Eds.), *Business-Driven Information Technology: Answers to 100 Critical Questions for Every Manager,* Stanford, Calif.: Stanford Business Books, 2003, pp. 269-273.

3. References to PLM did appear in the mid-1990s. However, it was unrelated to the current use of PLM. There were a number of computer suppliers that sold PLM as a program to manage desktop computers. In return for a monthly fee, these suppliers would install, maintain, replace (refresh), and dispose of desktop computers on a specified yearly cycle.

4. Source: B. Stackpole, "There's a New App in Town," *CIO Magazine,* May 15, 2003. The article uses as its source for the definition the University of Michigan PLM Development Consortium and ARC Advisory Group. Your author is a cofounder of the University of Michigan PLM Development Consortium and the developer of its PLM definition used by Ms. Stackpole. The University of Michigan PLM Development Consortium's definition is: "Product Lifecycle Management (PLM) is an integrated, information driven approach to all aspects of a product's life from its design inception, through its manufacture, deployment and maintenance, and culminating in its removal from service and final disposal."

5. The variations on PLM encapsulations continue to evolve. A newcomer is C2C or Cradle-to-Cradle, which reflects the focus on products that are not disposed of, but instead are recycled. See R. Smith, "Beyond Recycling: Manufacturers Embrace 'C2C' design," *Wall Street Journal,* March 3, 2005, p. B1.

6. See S.J. Fenves, R.D. Sriran, R. Sudarsan, and F. Wang, *A Product Information Modeling Framework for Product Lifecycle Management,* July 16-18, 2003. Paper presented at the International Symposium on Product Lifecycle Management, Bangalore, India.

7. See M. Burkett, K. O'Marah, and L. Carrillo, *CAD Versus ERP Versus PDM: How Best To Anchor a PLM Strategy?* AMR Research, 2003, p. 35.

8. See M. Iansiti, *Technology Integration: Making Critical Choices in a Dynamic World,* Boston, Mass.: Harvard Business School Press, 1998, for an excellent treatment of the NPD&I issue. While I do not dispute that researchers find that 80 percent of a product's costs are determined in the NPD&I phase, I would like to point out that this is based on how product costs are defined. If, for example, warranty costs or product liability costs are not allocated back to the product, but are allocated to, say, general and administrative, then the product costs will be understated. One $100 million dollar liability judgment can make a major percentage change in a product's cost.

9. Source: *The American Heritage® Dictionary of the English Language* (4th ed.), Copyright © 2000 by Houghton Mifflin Company. Published by Houghton Mifflin Company.

10. Source: Merriam Webster Online (www.merriam-webster.com/cgi-bin/dictionary).

11. At least it does in the well managed companies. See S.J. Spear, "Learning to Lead at Toyota," *Harvard Business Review,* 82(5), 2004, p. 78 for a description of how Toyota continues to experiment as it builds product.

12. This is true within any specific CAD application. There are issues when moving the math-based representations between CAD systems because different systems can and do use different mathematical techniques, which may not yield exact results between different systems. This issue has received a great deal of attention. The STEP specification is an attempt to develop a vendor-neutral specification that allows conversion between CAD systems. However, numerous issues remain, and the current solution to conversion problems is to stay within the native CAD system for assurance of consistency.

13. One division of the PLM software and services market is into authoring and analysis tools and collaborative product data management tools. Authoring tools are more or less synonymous with the CAD systems. For an example, see www.cimdata.com/press/PR04-0331.htm.

14. When drafting boards ruled the design centers, there were numerous techniques employed to reduce the amount of times a new drawing had to be done. Erasing, redrawing, annotations, and overlays were all techniques to reduce the amount of complete redraws. However, like the monks of the Middle Ages, a substantial amount of drafters' time was spent in copying from an old drawing to a new one.

15. The computer industry graveyard contains the remains of the many companies that developed specialized CAD hardware/software, for example, ComputerVision, Prime Computers, etc. Unigraphics, which is now UGS, Inc. appears to be the only company to have made the transition from proprietary hardware/software combinations to a software-only solution.

16. Because these technologies are interrelated, they are sometimes referred to by the family name of CAx. CAx refers to CAD (Computer Aided Design, CAM (Computer Aided Manufacturing), and CAE (Computer Aided Engineering). We deal with CAD and CAE in this chapter. We will address CAM later in the book.

17. See, for example, A. Saaksvuori and A. Immonen, *Product Lifecycle Management*, Heidelberg: Springer-Verlag, 2004.

18. Transactions are obviously not discarded. Completed transactions are used for reference and audit purposes. They are also used for data mining, where transactions are analyzed for trends and patterns.

Characteristics of PLM

T HERE ARE CERTAIN underlying characteristics that are an integral part of PLM. These characteristics need to be articulated for a robust understanding of the forces moving organizations to adopt PLM. These characteristics include

- Singularity
- Correspondence
- Cohesion
- Traceability
- Reflectivity
- Cued availability

We obtain these characteristics by examining an ideal model of PLM, the Information Mirroring Model, which captures the relationship between physical products and the data and information about those physical products. Before we do that, we need first to understand how that product information is organized today.

Information Silos

A common theme throughout the discussion of organizational structure is that organizations are divided into functional areas, such as engineering, manufacturing, sales, etc. The reason for this is as old as human activity. People are more productive when a task can be divided up into functional activities, and people can specialize in those respective functions.[1] However, there is a cost to this functional specialization. That cost is the development of silos of information where information is isolated and is principally contained in those same functional areas. While some of this information is localized and only of interest to the functional experts utilizing it, there is other information that is tremendously useful to other functional areas, but remains unshared.

As a result of this functionally oriented structure, the status of today's product information has the following characteristics. Information in this functionally oriented structure is siloed, ad hoc, duplicative, and inconsistent. We will explain what these various aspects of today's product information means below. With respect to information being siloed, if we look at Figure 3.1, we can get a visual idea about what we mean about siloing product information.

Figure 3.1 is meant to represent the state of the information systems within our current organizations. As noted above, information is generally organized by function. So such functional areas

Figure 3.1   Current Information Model

as design engineering, manufacturing engineering, sales and distribution, warranty and repair, and accounting all have information systems that reflect the needs of those functional areas. As with the silo-like water towers of Figure 3.1, these information systems contain the information and processes concerning product within each of these functional areas, and each of these silos has been built and operates independently of the others. A corporate-wide focus on product information is the exception, not the rule.

The difficulty arises when information from one functional area needs to be shared with another. Examples of this are plentiful. Design engineering has information about a product's shape and surface that manufacturing engineering requires in order to create the appropriate tool-cutting paths to manufacture it. Sales and marketing need information about the product's design in order to develop service manuals. Recyclers need information about product content in order to determine what part of the product can be recycled and what part of the product needs to be incinerated.

In the best case, the different functional areas package what is perceived to be the relevant information according to a preset procedure and formally hand this off to the next functional area. In a worst case scenario, the hand-off of information is done on an ad hoc basis, with the hand-off of information depending on the individuals involved, their working relationship, and the informal practices that have governed previous hand-offs. As illustrated in Figure 3.1, every hand-off requires a new "pipe" to be run between the functional areas.

In either of these cases, what was a flow of information within the functional area is choked to a trickle across functional boundaries. In addition to slowing down the flow of information to a trickle, useful and needed information is lost, diluted, and/or corrupted in the packaging process. The cost of re-creating or reconstituting information needed by the receiving area can be, and usually is, substantial.

In point of fact, a lot of these connections between functional areas are made fairly haphazardly. The information that is required and the information that is supplied occur in different forms and formats and are sent from one functional area to the other functional area in a manner that causes information leakage, the re-creation of information, and, in some cases, complete misinformation. It is not uncommon to find the recipient ignoring the packaged material because it is so unreliable and re-creating the needed information de novo.

Even when packaging is handled well, and hand-off procedures are in place, serious issues can occur over minor information transfers. On September 23, 1999, NASA's $125 million dollar Mars Climate Orbiter was lost because propulsion instructions were given to the spacecraft in English units of measurement, while the spacecraft systems were built to use metric units of measurement. This corruption in information flow, even though well packaged, caused the spacecraft to burn its propulsion system for longer than required and caused the spacecraft either to crash into Mars or carom off into space. Mismatched or corrupted information flow can be very costly even if the element of information seems to be relatively trivial.

This silo mentality also promotes suboptimal use of information. As functional areas work within their own silos of information, they tend to optimize their use of information within their own areas. However, this optimization on a local basis can, and often does, cause the suboptimization of information across the entire organization and even across the wider supply chain.

An example of this would be if design engineering determines that it does not need to develop information dealing with serviceability of the product. As a result, design engineering eliminates the resources that go into identifying, creating, and storing that information. However, this information may be a fundamental requirement to the group that services the product when it is in use. The cost to the service area of creating this information after the product has been developed and is being readied for the market may be significantly higher, sometimes by magnitudes, of the cost for which it could have been done in the design-engineering phase.

Despite this increased cost, which becomes evident by looking at the whole organization, design engineering is behaving in a perfectly rational fashion. The design engineers are trying to eliminate costs within their particular functional silo, but as a result of that, the entire organization suffers. This focus on optimizing in a specific functional area can have a substantially detrimental effect for the organization as a whole.

If we look at Figure 3.2, we see what the PLM Information Model is intended to be. Rather than the leaky substructure of Figure 3.1, what we have here is an organizationally created substructure that serves to identify and to connect all the functional areas to the product-centric data, information, and processes that the organization needs. Instead of each functional area being responsible for its own product information and processes, an organization-wide view is developed, and those product-centric data and processes are extracted and shared among the organizational processes. The result is an optimization of the use of product information across the entire organization and, taking a broader view, across the entire supply chain.

In addition, the connections between functional areas are not made on a one-off basis, but are made on a systematic and comprehensive basis across the entire organization. Therefore, other areas that may need to access that information, but either do not know the information is available or cannot afford the cost of creation within their functional area, can tap into this PLM information substructure and obtain substantial benefits from it.

The second aspect of today's product information is that a great deal of it is created and exists on an ad hoc basis. As mentioned

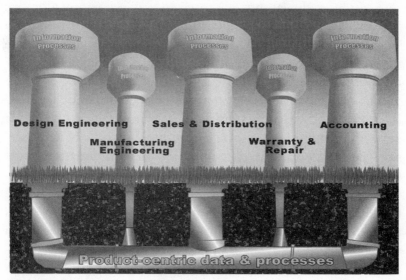

Figure 3.2   PLM Information Model

earlier in the book, the application where most engineering and product information resides is Excel. Excel is an extremely useful system on a personal basis, but results in ad hoc data structures because each engineer or group of engineers can create their own special worksheets to deal with specific information requirements. In some cases, these worksheets have similar, and, in point of fact, the same information as other worksheets, but in a form that is different and therefore not easily useable across the entire organization.

This results in the use of hundreds, and sometimes even thousands, of Excel spreadsheets that contain the same or similar information. However, this information that is used in the product development and manufacturing process exists as very separate formats and is not interchangeable across the entire organization—and sometimes not even interchangeable among subfunctions within the same functional area.

The cost of this to the organization is fairly substantial in terms of "hidden" information. This is information that has been created within the organization but can only be used for a specialized purpose that is defined by the use of the spreadsheets, and only by those individuals who understand how those spreadsheets are structured. Sometimes that is a group of one—the individual who has created that spreadsheet.

Because of this ad hoc approach to product information, we have the third aspect of today's product information environment: it is duplicative. What this means is that organizations have the same information in many, many different forms across not only the organization, but even within a functional area. For example, a group of engineers who keep information about specific characteristics of a product that they are working on together will have that same information in their own spreadsheets on each of their personal computers. While containing the same information, these spreadsheets may or may not have different formats.

As a result of this rampant duplication of information, the fourth aspect of today's product information is inconsistency. This is the characteristic that will cause by far the most problems for an organization. Inconsistent information leads to tremendous duplication of effort. This duplication of effort is an antithesis of lean thinking and a prime example of wasted time, energy, and material that information can and should be substituted for, with a concomitant savings in costs.

This is wasted time, energy, and material because engineers who are working with this inconsistent information wind up, in the best case, doing work that will have to be redone when they find that the information they have been relying on is not the correct version. In the worst-case scenario, these engineers will make decisions and implement actual product designs based on erroneous information. When those designs get into production and the product fails to work in the manner specified, the work will have to go back to ground zero and be redone using consistent information based on reconciling the various interpretations of it.

So it is the fact that these elements of today's product information are being siloed, ad hoc, duplicative, and inconsistent that causes a tremendous waste of time, energy, and material with a corresponding amount of additional cost and inefficiency for the organization. That is the reason that PLM, even when it is not focused on cross-functional uses, has substantial benefits within the functional areas because it reduces the waste of time, energy, and material that duplication and inconsistency of information causes. If we extend these issues across functional areas, the need for PLM becomes imperative.

Information Mirroring Model (IM Model)

The silo view of the information repositories and flows of our organizations is inefficient and suboptimizes our use of information. While this is true of the supply chain as a whole, our focus is first on our own organizations, over which we theoretically have a higher level of control. With that in mind, we need a different way to look at our product information. We need a way that focuses our attention on the product and not on the functions within the organization that produce it. This new view needs to help us understand how the physical product and all the information about that physical product relate. In addition, we need to understand how using the information about the product can allow us to substitute this information for the wasted expenditure of time, energy, and material—using bits instead of atoms.

In order to do all these things, we have developed the conceptual ideal for PLM that we call the Information Mirroring (IM) Model.[2] We use the basic idea of spaces in the IM Model, because when we think we instinctively think in terms of spaces. Even when the concepts we use are not spatial concepts, we think in terms of spaces. Examples of this abound. We aspire to *go up* the organization. We are concerned that profits are *down*. We are delighted that we are *under* budget. The supervisor is *over* the factory line workers. Finally, we have all had to deal with a *big* problem that we are glad to have *behind* us. In these and myriad other possible examples, we gravitate to spatial orientations when we think about both tangible and intangible concepts.[3]

Spaces are an especially comfortable way to think about products, because we naturally think about manipulating products in a spatial way. We view from different angles. We rotate products in different orientations. We explode the product into its component views. We color code temperature gradients. We show air flows across surfaces. Thinking about product information in terms of spaces is natural and useful.

The IM Model in Figure 3.3 is just such a model. It allows us to think differently about product information. Rather than being organized by functional areas, information is organized by product regardless of the functional area that may work with it. The IM Model is the implementation of the Info Core that we described in Figure 2.1 and the substructure in Figure 3.2. It is independent from the various functional areas, where different pieces of this information normally reside. The IM Model consists of four major components: real or physical space, virtual space, the linkages between real or physical space and virtual space, and virtual simulation spaces (VS_1, VS_2, VS_3,...$VS_{n,}$). We will describe each of these components in turn.

On the left side of the IM Model is real space or physical space. As human beings, we are very familiar with real space because it is basically the only space that we have worked with for the past 100,000+ years of human (or near-human) existence. We obviously have a very innate view of how we operate in real space since we do that from the day we are born. We learn very quickly the concepts and rules that govern real space—concepts and rules such as the

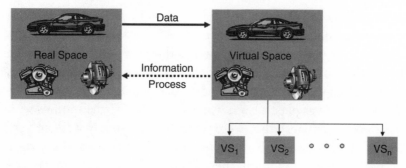

Figure 3.3 Information Mirroring Model

permanence of objects, the one-way direction of time, the existence of three dimensions, the pervasiveness of gravity, the requirement of energy to move matter. All these things are part of the concepts and rules of real space. As human beings, we all understand and naturally operate with them.

In a world of only real space, information has no separate existence from physical objects. If we want information about a physical object, we must be in close proximity to the object to sense the information about it. Its shape, weight, and other characteristics are only available if we can see and touch. In a primitive world inhabited only by quasi-amnesiac tribes with minimal short-term memory, there is no information without being in proximity to a physical object. The only way they knew how many and how big the wooly mammoths were was to be able to see them. The only way one quasi-amnesiac tribe knew they were at war with another quasi-amnesiac tribe was when a spear came thudding through their chests. An object and its associated information were inseparable.

On the right side of Figure 3.3, we have virtual space. As human beings, and unlike our quasi-amnesiac race, we also have always known virtual space. On an individual basis, we all have our own virtual space—our own minds. As discussed in Chapter 1, we all have to some extent the ability to capture information about real-world objects and virtualize them. Information about a physical object can be separated from that physical object and is accessible from our memory, even when the physical object is no longer available for inspection, although the accuracy of that recollection will differ from individual to individual and will generally decline with the complexity of the object.

However, for most of our existence this virtualization has been a very personal exercise. Our ability to share these virtualizations has been limited to impermanent media, such as sound waves or to static media such as paper. Even when we created virtualizations with motion (i.e., movies), they were static. Movies simply were and are static images shown one after the other. Once created, the images do not change. In addition, there is no independent processing of information in movies. Any information processing is still done in the human mind and then transferred to the celluloid medium.

As pointed out in the introduction, it has been only in the past 50 years, as computers have developed, that the ideal of shared, virtual space with independent information processing became possible. It is only within the last 10 years that this shared space became universally accessible through the explosive expansion and use of the Internet.

The creation of shared, virtual space has had the opportunity to change radically the way we view and use information. The implementation of a shared, universally accessible virtual space has changed the dynamics and practice of how we deal with information about physical objects.

Prior to this shared, universally accessible space, the only practical way to have information about a physical entity was to examine the physical entity itself. Much like our quasi-amnesiac tribe members, if we wanted to access the information about a physical object accurately, we needed to be in proximity to that object. While, in theory, we virtualized the information on paper records, these paper records were prone to not being updated to reflect changes. Even if they were, they existed in a distinct geographical location that more often than not was not the geographic location where they were needed.

However, even being in close proximity to the physical object might not meet our informational needs. If the information required was not accessible to our senses or instrumentation, we would have to expend time, energy, and material to make it so. If we needed to know the specific model of a fuel pump deep in a helicopter's engine, then we would have to expend time, energy, and material to disassemble, inspect, and reassemble that engine.

Even when the information is not hidden, if we wanted to know what tires were on the car in Figure 3.3, we went and looked at those tires. If we wanted to know the type of motor that was in that car, we had to go look in the car itself. Before computers, driving into the service bay of an automotive dealership, even the dealership that sold us the car, was an event that our quasi-amnesiac tribe could appreciate. Every visit was as if this were the first time the dealership had seen the car, especially if we had it serviced at other dealerships or service garages in between visits. The service representative asked us what we were there for, and, if the mechanics wanted to know what parts were in the vehicle, they opened the hood to look.

If all we could do was to access the information in virtual space about physical objects without having to be in proximity to or disassemble that physical object, it would be highly useful. But we can do more than that. Because of the processing power of computers, we can derive other useful information from the information we have about the physical object. Not only can we can disassemble the physical object virtually, but we can take cross-sectional or other hidden views that would only be possible in real space by destroying the actual physical object.

We can rotate the object in virtual space and examine it from all sides. We can see if its movement interferes with another physical object with which it must work. With enough information about the physical object and some basic geometric manipulations, we can meet the Grieves Visual Test of Virtualization. For simple objects, we are almost there. For more complicated objects, we are not far from having that capability.

If the requirement for PLM were simply to strip the information from a physical object at a certain point in time and maintain it for later inspection, then we would be close to achieving that goal. Or if the requirement were simply to go the other way and create a physical product from our virtual specifications, then we would be close to achieving this goal also. However, as we concluded in examining the definition of PLM in Chapter 2, these are only part of the goals that PLM strives to meet.

PLM is also tasked with maintaining the linkage between the physical object or the product and its related information throughout the product's life. It is not even close to sufficient to capture this linkage at a particular point in time. There must be a mechanism to maintain and update this linkage throughout the life of the product. This is the role of the third component of the IM Model, the linkage between real space and virtual space.

Real space and virtual space will have to be linked in two ways. The first way is from real space to virtual space. As information and data are developed in real space as the state of the product changes, then virtual space will need to be updated with that information. For virtual space to substitute atoms with bits, virtual space will need to "mirror" the change in product information in real space. This mirroring of information is shown with the arrow labeled "Data" that goes from real space to virtual space.

Conversely, in real space we will want to use the information that virtual space has. Taking our example of the helicopter fuel pump, which would use less time, energy, and material? Tracking down a fleet of helicopters based all over the world, sending a mechanic to each of those locations, and disassembling each helicopter engine to inspect the fuel pump? Or querying our computer system and sending our traveling mechanic only to those helicopters that have the fuel pump in question? The answer is obvious.

In other aspects of the product's life such as recycling and disposal, the processes that we currently build in at the front end are a wasted effort without PLM. We have Design for the Environment (DfE) initiatives that allow us to design the product to be recyclable at the beginning of the product's life cycle, Yet when the product is taken out of service 10, 15, 30 or even more years from now, the knowledge about that recyclability will more than likely have been lost. Both information and processes must be linked to the product and then must be available in real space when we need them.

The fourth component of our IM Model is the virtual simulation spaces ($VS_1, VS_2, VS_3, \ldots VS_n$). A major advantage in virtual space that we don't have in real space is the property that, as shown on those little blocks underneath virtual space, we can have sub-virtual spaces. This ability to run simulations may be the most powerful aspect of PLM in the future, and one we will cover in more depth later.

What this means is that we can run simulations of processes in virtual spaces that we would normally have to actually perform in real space. In the digital factory environment, we can build a factory, put the robots that do the welding in place, and run virtual products through that space such that we can see that the processes we have put into place actually work in our simulated operations. In point of fact, having as many simulated spaces as we can possibly have, we can run any number of simulations and various factory layouts until we get one that meets the requirements that we are trying to obtain.

The drawback of developing physical factories, even small pilot versions, versus developing digital factories is that, while we can do a fair amount of planning on the front end, once we decide to build the physical plant, the plant is built. We will then have to work out the problems with production by utilizing real equipment, real people, real energy. And the cost of that is fairly substantial compared to its simulated equivalent.

The Information Mirroring Model not only allows us to capture and represent information as we move along in the product's life, but also allows us to simulate various actions to the product that would be prohibitively costly, if not destructive, in real life.

Characteristics of PLM

Based on examining the IM Model, what characteristics should PLM have? In order to realize the model contained in Figure 3.3, Product Lifecycle Management information should have the following characteristics, as mentioned at the start of this chapter:

- Singularity
- Correspondence
- Cohesion
- Traceability
- Reflectiveness
- Cued availability

What do we mean by these characteristics? What we mean by these characteristics is that Product Lifecycle Management systems will need to have these aspects in order to fully represent the information that they contain and present it to the end user.

Singularity

Singularity is one of the most important characteristics of PLM and might be one of the most elusive. Singularity within PLM is defined as having one unique and controlling version of the product data. While the term *unique* is self-evident, what we mean by controlling is that when we have two or more unique data representations, one of the representations is the one we all agree is the correct one—the representation that everyone will work with.[4]

Product data singularity has always been an issue. (Og thinks the clan's inventory of spears is five, while Ug thinks it's six.). Singularity of data is a relatively recent problem as it pertains to formal designs and plans because the more complex and voluminous the designs and plans are, the more difficult and costly it is to duplicate them. In addition, when the designs and plans had to take physical form, such as drawings on paper, we could always resort to geographic coordinates to uniquely identify the data and its various representations. Given its size and complexity, there most likely was only one set of drawings and plans for the Eiffel Tower. However, even if there were multiple sets, the drawings hanging on the walls of Gustave Eiffel's office were the controlling set.

The development and use of computers compounded this problem of lack of singularity because now even the most complex and voluminous data could be duplicated at minimal cost and effort. Now there can easily be multiple copies of the most complex product data. Or more to the point with respect to the issue of singularity, there can easily be multiple copies of the most complex product data that varies slightly with each copy because of small but significant changes that have been made between each copy.

There are two cases we must examine. The first case is when the product data represents an existing, tangible object—a part, a component, an assembly, a product. The second case concerns the situation when the product data represents what is not yet an existing, tangible object.

The first case, the case of the existing tangible object, is the easier case to deal with. When there is a tangible object, we can use it as the reference to determine which of the product data is controlling. By definition, the data about a tangible object is contained in the object itself. If there is some question about the accuracy of our product data concerning the shape of this existing, tangible object (i.e., its dimensions, weight, and its color), we can always measure the dimensions of the object, weigh it, and take a color spectrometry of it to compare our product data with these observations. When the spear product data inventory count of Og and Ug disagreed, they could walk back into the cave and do a physical count to find out whose inventory record should be the controlling one.

The second case, where there is no tangible object to use as a reference, is more problematic. This is the very common situation where a component or part is in the design stage and there is as yet no physical representation of it. In this case, there has to be a shared understanding by the individuals involved as to how to identify the controlling product data if multiple unique versions exist.

When this design was embodied in a paper drawing, the drawing could and did serve as the tangible reference. Like Gustave Eiffel's tower design, the paper drawing was hung on the wall. If there were any questions as to which version was the controlling version, one simply walked into the room and looked at the drawing on the wall. When organizations were small and performed all the design functions in a single location, the "drawing on the wall" technique worked reasonably well.

However, even before computers proliferated digital copies, as organizations grew in size and expanded to other locations, the "drawing on the wall" model began deteriorating. Now there were unique and different drawings on other walls. This led to the all too common occurrence of design teams finding out that the substantial number of worker-hours spent on a particular design were wasted because the drawing they used as their basis was not the controlling revision.

The world of digital representation made this problem worse. The ease of making digital changes meant that the product data could change almost moment by moment, and the ease of making copies that were exact duplicates only for a brief period of time meant that more individuals could be making those changes. A new task in the design process—reconciling work done on different versions of product data—became a standard practice and the source of significant wasted worker-hours.

The lack of information singularity is a major source of wasted time, energy, and material. It is a common occurrence to discover that the product plans that our engineers are working with to make their designs are not the current revision and that the work will have to be redone. This occurs regularly within functional areas such as the design function where engineers working on the product are using different information. As we cross functional areas, say from engineering to manufacturing, the issue is more likely to occur than not.

Software applications that implement PLM must have as a fundamental characteristic the ability to manage this singularity of product data. PLM systems must have the ability to identify the controlling product data so that there is no question, if there are multiple versions, which is the one that everyone refers to. For product data in the design phase where there is no tangible object to serve as a physical reference, this is a fundamental purpose of PLM. The alternative is to expend time, energy, and material in reconciling different versions of the product data and redoing the work that cannot be reconciled.

However, this is highly useful even for product data where there is a tangible object to serve as a reference. There is a cost to collecting data from the tangible object, the time, energy, and sometimes material to disassemble, measure, weigh, etc. If the tangible object is unchanging, then the time, energy, and material used in collecting this data and identifying the correct product data version is wasted.

At its simplest, PLM implements a mechanism whereby there is a unique reference to the product data. If someone wishes to work with the product data it is "checked-out," worked on, and the revised version "checked" back in. The actual software implementations vary. Since we are talking about digital data, nothing is really "checked" in or out. In some cases, the data is copied to another location (i.e. another computer file in the same or another computer), worked on, and the revised data copied over the original data. In other implementations, the product data will be revised within that data file, but will be unavailable to anyone until it is checked back in.

The description of this mechanism is meant to illustrate PLM's ability to implement data singularity. The larger the granularity required to be checked out, the larger the amount of data that is locked out for use by other collaborators, and the less concurrent work that can be done. We will later explore other mechanisms that need to be pursued in order to reduce this limitation.

Before we leave the topic of singularity, we must make the point that there is nothing inherently wrong with duplicate copies of information. There are invaluable uses for duplicate information, such as back-ups that insure that information will not be lost—a major issue with destroyable paper-based information. Duplicate information is also useful for experimenting with alternate designs on a local, temporary basis. It is not knowing which data is the controlling data and which data is the duplicate that is the issue.

We do not expect to get singularity in the relatively near future, but every time we can reduce an instance of the same information being duplicated in various systems, we will get closer to our goal of a singular version of data. Doing so will increase productivity because there will be less waste of time, energy, and material from working with wrong product data.

Correspondence

Another key characteristic of PLM is correspondence. Correspondence refers to the tight linkage between a physical object (e.g., a component, part, or product) and the data and information about that physical object. The data regarding geometrical shape and dimensions and characteristics such as material, weight, color, tensile strength, etc. is embodied in physical objects through the type and arrangement of the atoms that make up that object. The task of getting these physical objects to give up that data is developing the methodology and technology required to extract the data from the physical object. Measuring devices such as micrometers and laser measuring systems, coordinate measuring systems, scales, spectrometers, and X-rays, among other technologies, were all invented and developed to allow us to extract data and information from the physical object in which they are intrinsically intertwined.

When we are the ones creating the physical object, the issue of the data is easier to deal with. We define the geometric shape and material. If we have our physics in good shape, we also define the weight, color, and tensile strength of the physical object we wish to create and build. Here, the methodology and technology are about making the atoms conform to the data, not extracting the data from the atoms. Chemical processes, milling machines, presses, welding machines, and other atom forming or shaping technologies help us realize the data in physical materials.

Regardless of whether we have the physical object first and extract the information about it or whether we have the information first and create the physical object from that information, a core characteristic of PLM is developing and maintaining a correspondence between the physical object and the information about the physical object. There are a number of significant reasons to do so.

The first reason is our interest in replacing wasted time, energy, and material with information. If we do not separate and maintain the information about our physical object, then any time we want the information about it we must expend time, energy, and material to get it. We must spend time and energy to locate and possibly move the physical object. We must expend time, energy, and material to obtain the data and information through the appropriate measurement systems. To compound matters, there are sometimes measurements that we can take only by destructive means.

Let us take the example of the simplest of all measurements, binary counting. We want to know whether our particular model fuel pump is or is not in the engine of each helicopter in a fleet of helicopters. If we do not have correspondence between the helicopters and the data, in this case the parts that make up the helicopter, we have to track down each helicopter, send out a mechanic to open up and disassemble each engine, and make the determination whether or not the part in question is present.

If the correspondence between the helicopter and its parts is maintained, then all we have to do is request a list of the helicopters that have the part in question. If the purpose of this exercise is to replace a defective part, then regardless of whether there is correspondence or not, the cost of replacing it on the helicopters that have that part is the same. But the savings in time, energy, and material in not locating, opening, and disassembling the engines of the helicopters that do not have the part in question is tremendous.

Second, if we do not maintain this correspondence between the data and information about the physical object and the physical object itself, then the only way that we can obtain that data and information is by actually possessing the physical object. But this is something we have gotten away from. If we walked into the office of the chief engineer 30 years ago, we would see an office filled with parts. This was the informational database. He or she did not need correspondence—the physical object itself was present. However, if he or she wanted to acquire or reacquire information about the physical object, the chief engineer had to expend time and energy to do so. It had to be sent it back to the lab to be disassembled, measured, weighed, etc.

If we walk into the chief engineer's office today, the chances are that we will find a computer screen in his or her office, but very few actual parts. If the correspondence between a physical object and its informational description is corrupted, lost, or did not exist in the first place, the ability to know about that physical object is greatly diminished.

Parts reuse is to a great extent driven by correspondence. If correspondence does not exist, then an engineer is going to design a new part instead of using an existing part. If the information about a part does not exist, then the only way an engineer knows it exists is having access to the actual part itself, something that is becoming increasingly rare as engineering goes global and manufacturing takes place far from the design engineers.

Cohesion

The next element of PLM information characteristics is cohesion. By cohesion, we refer to the fact that there are going to be different representations or views of product information depending on our perspective of the product. Since we are discussing tangible products, we always have a geometrical, or what is often referred to as a mechanical, view that reflects the physical structure of the product. This mechanical view is a geometrical, three-dimensional view of the product. It shows the form of the product as we see it in three-dimensional space. Views of hidden surfaces and cross-sections will by definition be cohesive, because these views are always derivable from the geometric definitions.

However, in addition to this geometrical or mechanical view, we may also have other views of the product that we abstract in order to focus on what functions are performed by these views or aspects of the overall product. We do this because we want to focus on a certain aspect of product functionality without being overwhelmed with other aspects that are not relevant to our interest. We have created this technique of views to compensate for our constrained human memory and processing capability. Within this view, we are concerned with the logic of of the view, not the physics of it.

For example, many products have an electrical view, which is the logic diagram or schematic that shows the electrical system contained within the product. The length of the visual lines connecting electrical components has no relationship to the length of the actual wires that are required to physically connect these components within the product. In addition, what looks like discrete components on the logic diagram or schematic may be embodied on a single chip. The reason the logic diagrams visually look like this is because this representation is for the purpose of understanding how the electrical system logically functions, not how it is physically realized.

But this logical view of the electrical system must eventually be represented in physical components. Wires must be run. Specific chips must either be selected or designed. Voltage sources must be bought and installed within the product. If the electrical schematic changes because someone realizes it will not produce the desired function, then the physical implementation also must change. If a physical component changes that will change the functionality of the electrical system, then the electrical view must change to reflect that. This is cohesiveness.

The physical product obviously has the superior position. We can change our logic diagram as much as we want to produce certain functionality. However, if we do not implement it with the requisite components and wiring, it will not make a bit of difference in the physical product. The functionality of the physical product is determined by atoms, not bits.

However, if we are going to mirror the product's functionality in virtual space, we are going to need to have this cohesion between views. While cohesion is not a problem in real space since there is only one view of the product, the actual product itself, it is a problem in virtual space. In real space, we have a product that has the information about it as part of its very makeup.

In virtual space, we have attempted to create a representation of the product by consolidating its different views in different computer programs: one for the geometrical representation, one for the electrical schematic representation, one for the hydraulic system representation, one for a BOM view, etc. Rarely are these views brought together.

Not to do so is a cause for wasting time, energy, and material. If we work with non-cohesive views, we run the risk of thinking we have a specific functionality because it is specified in our abstracted diagram, but it is not being implemented in the product because the actual components are inconsistent with the logic diagram.

Cohesiveness has not only been a problem between views of different perspective, but cohesiveness has been a problem between views of the same perspective. A common problem in most organizations is a lack of cohesiveness in the Bill of Materials in different functions. Engineering has one BOM. Manufacturing has a different BOM. Finance has a BOM that matches neither Engineering's nor Manufacturing's. As a result, inconsistent and costly decisions are made by all three parties.

The issue of cohesiveness is becoming more complicated. In this day and age, we more often than not have a programmatic or software view. This is the view of the product that contains the exact software program that resides in the component. Because functionality is often enabled by programming rather than discrete circuitry in today's products, the same basic circuitry may have different software associated with it, depending on the desired functionality. Small, easily made changes in software may have major implications for how the product functions—or does not function. The maintenance of this programmatic view has become such a major issue that it has spawned it own discipline, configuration management (CM), in an attempt to institute discipline and control, However, configuration management is also a part of cohesiveness.

One approach to obtaining cohesiveness is to reduce the number of independent views and derive the abstracted views from a limited number of richer views. For example, the selection of geometric points or the mesh that is needed for a view in Finite Element Analysis (FEA) used to be independent of the CAD representations.

The engineers responsible for FEA would select their data points from the CAD drawings or even from measurements of a physical mock-up. They would run their stress analysis. If they changed the mesh to get a better result, they needed to inform the design engineers and the CAD designers so both of these could update their information. This often was not done or not done in a timely fashion, so there was a waste of time, energy, and material as views had to be reconciled and products scrapped.

The state of the art is now to derive the mesh from the CAD representation. If the engineers make changes to the mesh to produce better results, then these changes are automatically fed back to the CAD representations in order to change the geometry of the product. This has improved the cohesiveness of these different views. There are other abstract views, such as BOM views, that similarly could be at least partially derived from the geometric view.

Moving in the direction of fewer independent views implies that when we obtain a particular view of that product, we will get all the other views that will be consistent with the way that product is structured and manufactured. While we can get to the Grieves Visual Test of Virtuality without cohesiveness, we cannot get to the Performance Test of Virtuality. Without cohesiveness, we cannot be assured that when we run specific tests like crash tests or circuitry tests in order to determine functionality that these tests will accurately reflect what will happen in an actual product under the same conditions. It is the ability to run these tests in virtual space and not in real space that will help drive the next wave of productivity.

Traceability

Traceability is the ability to demonstrate that the path of a product's travel through time can be followed seamlessly back to its origin. The reason that traceability is important is that, unlike our semi-amnesiac tribe, we do not start every day anew, but instead rely on what we have done the day before to continually progress. However, we usually have done a number of things the day before, some successful, some not so successful. We need to know that we are relying on the successful things we did in order to progress.

In addition, we continually put our ideas, designs, and product to tests to determine whether or not our assumptions about the attributes and functionality of our ideas, designs, and products are disproved.[5] However, we can rely on these tests only in two situations. The first is if we are referencing the unchanged version we tested. The second is if we are referencing derivatives of that version where we have high confidence that we can understand how the changes affect the attributes and functionality and where we can compute with a required degree of confidence new test results that would mirror the actual tests if we were to perform them. If we cannot have this required degree of confidence, then we have to run new tests.

The result of a lack of traceability is that we run the risk of wasting time, energy, and material in a number of different ways. First, if we cannot determine which the successful version was and which the unsuccessful version of our previous efforts was, and, if we chose an unsuccessful version to build upon, we would eventually realize, after wasting time, energy, and material, that we chose the wrong one. We will then have to go back to our decision point, attempt to understand which the correct choice should have been, and start all over again.

Everyone in design and development can tell at least one story where the wrong design or prototype was selected for further revisions. They worked on it fruitlessly until they figured out that this version could not have been the right version. It was literally "back to the drawing boards" to find the right version and start all over again. The cost? Wasted time, energy, and material.

We addressed traceability in the physical world by creating physical evidence of this traceability—namely documentation. We created separate pieces of paper. We organized the documentation so that all the material, designs, notes, drawings, and tests that were related to a specific version were collected together. We then ordered those pieces of paper in chronological order so that we, in theory, could follow the path back to its origin.

The key words here are *in theory*. In the physical world, this documentation process is a human undertaking, and the documentation process is generally not high on the list of desirable activities for human beings. We would rather create and do than document. As a result, this "in theory" documentation often was done sketchily, done long after the fact, or not done at all. In addition, documentation has a nasty habit of getting misplaced, lost, or destroyed. Documentation is costly to produce and equally costly to provide at the right place and time.

This is why everyone who is in design and development can tell their story of wasted effort. In fact, one does not have to be in design and development. Anyone who has had to rely on documentation can tell a similar story of wasted effort when the documentation did not reflect the actual product.

In some products the lack of traceability is simply an annoyance and a waste of time, energy, and material. In other products, it is a matter of life and death. In the aerospace, medical device, and the automotive industries, to name three, traceability is both a legal and moral requirement. In these and similar industries, there cannot be any question as to whether the product's life can be traced seamlessly back to its origin or what version the testing applies to.

It is a core requirement, and failure to be able to do so invariably carries criminal penalties and sanctions. It would be unconscionable for a medical device manufacturer to say, "We tested version A. We then made substantial changes to make version B. We don't need to test version B because it should be good." The CEO of that company would only be seeing his or her family on visiting day for a long time to come.

In industries where human safety is an issue, any change to the product, no matter how trivial it may seem, triggers a requirement for new testing before the new version is approved. In certain industries, such as aerospace, changed parts are required to go through a process called *effectivity*, whereby parts must go through a testing, validation, and approval process before they can become effective, and they are given an *effectivity* date when they can legally be used in production.

In these situations, when human life is at stake, information necessary to replace the wasted time, energy, and material must be obtained. The calculation of whether the cost of that information is less than the wasted time, energy, and material is not done, because in this case the "wasted material" is human life. Traceability is integral to the industries where human life is so closely tied to product performance.

The move into the virtual world helps to remedy the issue of traceability and, at the same time, raises new issues. On the remedy side, once information is moved into virtual space, it can easily be backed-up so that it never can be destroyed. In addition, traceability can be built into the process of creation so that documentation is not an after-the-fact proposition, but is a by-product of the creation process. In addition, processes such as effectivity can be built in so that the premature release of a modified part is not possible. Finally, with the right search capabilities, we can be fairly well assured that documentation can always be located and available any time, anywhere.

The new issues that a move into the virtual world raises are around the lack of physical documentation. We no longer have or want separate pieces of paper that provide the time-honored "paper trail." Since information can readily be changed in virtual space, we will need reliable mechanisms for knowing when and where the information was captured and that it has been unchanged since that point in time.

The move into virtual space should greatly improve traceability. The ability to capture information at its source and the ability to find that information when needed will enhance the traceability of products, both in development and in use. PLM must have traceability as a core element.

Reflectiveness

Reflectiveness is directly related to the arrow in the Information Mirroring Model that connects the real space to virtual space and captures data and information from real space into virtual space. In real space, when we change the state of anything, trim a little material off a part, assemble two parts together, erase one line and draw another on a piece of paper, the information changes because it is intrinsically part of the atoms that are impacted by those changes. If we are going to separate that information and create an image of it in virtual space, then we need a mechanism to change the information in virtual space when the corresponding information changes in real space.

In the same way that the image in a mirror reflects changes in real space simultaneously with changes that occur to physical objects, so the ideal of PLM captures those changes in virtual space. In the same fashion that we can make decisions about the state of a physical object by looking at its image in a mirror and knowing that there is no lag time between the changes that occur and the image we see and no loss of detail between the physical object and its image in the mirror, we would like to be able to rely on PLM to provide us with similar timeliness and fidelity of information.

We use the word "similar" for a number of reasons. First, the analogy with a mirror has its limitations. A mirror is relatively inexpensive because its images are transient and powered by light, which is cheap and plentiful. We want to capture and preserve the information. As mentioned earlier, capturing and processing information is not costless and, depending on the technologies involved, may be exceptionally expensive.

Second, we want to capture more than surface images. That generally takes instrumentation to collect and communicate. Again, this may not be an inexpensive proposition. We will need to calculate the cost of collecting and communicating whatever information we need or want against the value of that information. This will determine the timeliness and fidelity of the information. To decrease costs in order to bring this calculation into balance, decisions will be made to lag or batch the timeliness of the data collection and select only some of the information to be collected and communicated into virtual space. In the majority of cases, the lag in time or diminishment in fidelity will have little or no impact on the decisions made from the information in virtual space.

The whole point of reflectiveness is to allow us to substitute this information for wasted time, energy, and material. If we can examine virtual space when we need information, it will be substantially less costly and time consuming than examining real space. What takes less physical effort and time? Accessing the inventory record that contains the quantity and location of a specific item about which we want to know how many and where they are located, or sending people out into a number of warehouses spread out across the country to locate the same information about that item?

Which is more efficient is not even close. However, this is if and only if the virtual information about the inventory is a reflection of the inventory in the physical world. If it is not, we are better off physically searching the warehouses, because the virtual information cannot be relied upon.

To the extent that we work more and more in the virtual world, reflectiveness gets easier. When designing was done on paper, we had to expend resources to move that information from the physical paper to the virtual memory of a computer system. With the proliferation of software for design, the issue of reflectiveness becomes moot. It is only when the design takes form as a physical product and becomes functional that reflectiveness matters.

The real value of reflectiveness will become apparent when we begin to track "as-built" products. Today, when we want to know if a specific component is in a specific product, we generally resort to logical deduction: "This serial number was built on such-and-such a day, and, based on that date, we were using components from this batch based on when we received that batch and the rate with which we built products." At best, this is a time-consuming approach to the issue, and at worst the logical deduction is wrong. Capturing as-built information and updating this information as the product progresses throughout its life ("as-maintained") is an integral benefit of PLM. There are issues, however, with privacy and security that we will touch on later in the book.

Cued Availability

If reflectiveness is related to the arrow indicating data movement from real to virtual space, then cued availability is related to the arrow indicating the movement of information and processes from virtual space to real space. Cued availability is simply being able to have the right information and processes when we need them. The term *cued* indicates that we might or might not be searching for this information and these processes, but because of the situation, we are presented with them.

Without cued availability, the Information Mirroring Model is simply a historical vault. It would collect the changes that took place in the physical world, but would provide limited utility for that collected data and information. In the worst-case situation, the data would be collected in virtual space, never to be seen again. A slightly better situation would be that the collected data and information would only be useful only as a forensic tool in a post-mortem of what went wrong.

This was the situation in the famous Ford-Firestone debacle, where there was a rash of rollover accidents involving Ford Explorers equipped with Firestone Wilderness AT tires. That resulted in more than 174 deaths. While the debate still continues as to whether the root problem was with the SUV or with the tires, the information that there was a developing problem was buried in the warranty reports collected by Ford. However, it was not until after the deaths occurred and potential damages in the hundreds of millions, if not billions, loomed that the warranty data was analyzed.

For information to be worth the value of its capture, processing, and storage, it has to replace wasted time, energy, and material, not confirm that the time, energy, and material was indeed wasted. While there is some value to forensic use of information, the real value of information is preventing the waste of time, energy, and material.

We are rapidly improving our capability to search for information in virtual space. It was not very long ago that we could only realistically search structured information in predefined and managed databases. The Internet gave us access to volumes of unstructured data and information. Search engines such as Alta Vista and Lycos were created and set about discovering and cataloging the unstructured information on the Internet. Google is but the latest of these search engine creations that easily and effectively search the unstructured information on the Internet, and these same technologies are brought to bear on the nonpublic information that exists within organizations.

We are rapidly approaching the time when, if the information exists in virtual space, we are able to search and find it. That is one aspect of cued availability. However, fully cued availability requires something more. It requires that the information is presented to us when we may not be searching for it, but need it nonetheless.

An example of this is that we spend a great deal of resources on Design for the Environment initiatives, where we expend resources in the design phase to develop the processes to recycle the product at the end of its life. The issue today is that without PLM it is hard to understand how the recycling center years, if not decades, from now will know those processes exist, let alone search for them. Cued availability implies that when the recycler accesses the product information in virtual space, he or she will be presented with the appropriate information and processes.

Developing the appropriate triggers and cues will be a focus of PLM in the future. In addition, there is ongoing work in having search technologies determine meaning and relationships not just search by specific words. True cued availability will depend on the progress in this area.[6]

Summary

In this chapter, we examined the current state of product information, which is best described as siloed. Product information is organized in silos by functional area, and sharing information across functional areas is an inefficient and duplicative process—if it occurs at all. We proposed a new way of thinking about product information, the Information Mirroring Model, which creates and mirrors a virtual representation of physical products. From that model we derived the characteristics that PLM will need to have. Those characteristics are singularity, cohesiveness, correspondence, traceability, reflectiveness, and cued availability.

Notes

1. While the concept of division of labor extends back into prerecorded time, the articulation of this concept is generally credited to Adam Smith and his contrast of the productivity of pin-makers who make the entire pin versus those pin-makers who divide the task into "eighteen distinct operations" with different individuals specializing in one or two of those operations. See Book I, Chapter 1 of A. Smith and A.H. Jenkins, *Adam Smith Today: An Inquiry into the Nature and Causes of the Wealth of Nations*, New York: R.R. Smith, 1948.
2. The Information Mirroring Model has been a foundational view of PLM for me. I have used it from my early discussions and presentations on PLM because I think it captures the larger view about the representation of information.

I have evolved the model and probably will continue to do so. A description of the Information Mirroring Model appears in M. Grieves, "Product Lifecycle Management: The New Paradigm for Enterprises," *International Journal of Product Development*, 2(Nos. 1/2), 2005, 71-84. The description in this chapter is an evolved version of that contained in the paper.

3. This goes to the heart of the way we think. One explanation was that these uses were examples of "dead" metaphors, metaphors that were simply used without being connected to any spatial reference. However, the latest and most persuasive research is that our brains are wired so as to use sensorimotor areas of the brain in conceptual thinking processes. Thus these metaphors are not "dead" at all, but very much alive and highly useful when thinking about conceptual matters. See G. Lakoff and M. Johnson, *Metaphors We Live By*, Chicago: University of Chicago Press, 1980; G. Lakoff and M. Johnson, *Philosophy in the Flesh: The Embodied Mind and Its Challenge to Western Thought*, New York: Basic Books,1999; and, for a slightly contrary view, J.R. Searle, "Metaphor," in A. Ortony (Ed.), *Metaphor and Thought* (2nd ed.), pp. 83-111. New York: Cambridge University Press, 1993. For a view of how spatial thinking extends even to formal, legal documents see M.W. Grieves, *Business Is War: An Investigation into Metaphor Use in Internet and Non-Internet IPOS*, Unpublished EDM diss., Case Western Reserve University, Cleveland, 2000.

4. Some will refer to the controlling data representation as the "true" one. I am reluctant to use the word *true* in this context. It is not *true* in any positivist, independently verifiable sense. It is only *true* in a social constructionist sense that the participants agree. "Controlling" is a much less confusing description.

5. As Karl Popper pointed out, we can never prove anything, we can only disprove. See K.R. Popper, *The Logic of Scientific Discovery*, New York: Basic Books, 1959.

6. For a good overview article of the direction that this should take, see T. Berners-Lee, J. Hendler, and O. Lassila, "The Semantic Web," *Scientific American*, May 2001, 34-43.

The Environment Driving PLM

W HAT IS DRIVING the need for PLM? Why aren't the previous techniques and technologies sufficient for today's organizations? This chapter brings these issues into stark relief by comparing the changes in organizations and their environments over the last 30 years. While the changes appear incremental when looked at from day to day, the differences are truly dramatic when viewed over a longer time frame. Chapter 4 explores how scale, complexity, cycle times, globalization, and the regulatory environment are changing the way organizations need to deal with product-based information.

This chapter will also explore the fundamental requirements emerging from within the business environment that are driving PLM. These requirements include a need to improve productivity, the rate of innovation, collaboration, and quality. Finally, we discuss the fact that the ultimate driver of PLM and the boardroom decision to invest in PLM solutions will be their ability to create value for the organization. To conclude, Chapter 4 will look at how that assessment is made by introducing the IT Value Map and comparing PLM to other information technology (IT) initiatives.

External Drivers
Scale

In order to appreciate the scale of today's organizations, we need to go back and look at an extended time period to see what changes, if any, have occurred in our organizations over that period of time. The issue we generally face as human beings is that it is sometimes difficult to see change occur as we live it from day to day. Like the proverbial frog that never realizes it is in boiling water because the water temperature has increased so gradually, we often do not realize the major changes that have taken place because of their constant, but gradual, occurrence.

It is only when we take a longer view that the elements of change and the magnitude of change come into sharp focus. If we look at the past 30 years, it is difficult not to see the immense changes in scale that have occurred in our organizations. While we have always viewed our large corporations as being large within our society, we really don't get a handle on the size change unless we do a comparison over the past 30 years.

Let's take a look at some of these companies and see the change in scale that has occurred during this period of time. We have four companies to look at: General Motors, Ford, General Electric, and Wal-Mart. There is nothing particularly unusual about these companies, and we could have selected others that would have illustrated the same characteristics of growth.

Table 4.1 shows these four companies along with their annual sales in 1973 and their annual sales in the year 2003. General Motors had revenues of $35 billion in 1973. In 2003, GM revenues grew to $186 billion. Ford Motor Company, also in the area of automotive sales, was $23 billion in 1973 sales, and, in the year 2003, was $164 billion. This is more than a fivefold and a sevenfold increase in sales, respectively—so much for downsizing.

General Electric, a company in a substantially different area as well as a company that has embraced the service industry through its financial arm, had sales of $12 billion in 1973, and in 2003 had annual sales of $134 billion. However, the all-time leader on this list is the Wal-Mart Corporation, which had 1973 sales of $126 million and 2003 annual sales of $255 billion; a 2,000 times increase during that period of time!

Table 4.1 Scale

Company	1973	2003	Δ
General Motors	$35B	$186B	5.3x
Ford	$23B	$164B	7.1x
General Electric	$12B	$134B	11.1x
Wal-Mart	$126M	$255B	2,031x

Source: Annual Reports, 1973, 2003.

It is hard to come up with a reason other than the use of information systems that would have allowed these companies to grow in such magnitude during such a short period of time. It would be hard to attribute such growth to organizational efficiency, the brilliance of the current crop of executives (as opposed to the executives of the past), new methods of production, or anything else other than the rise of information systems since the 1960s.

If we were to hypothesize that information systems enabled this growth, what our hypothesis would state is that the simpler the products of the organization, the more information systems would be of use. With simpler products, information systems are more effective in replacing wasted time, energy, and material with information. That premise seems to hold true because, in the case of Wal-Mart, the information that Wal-Mart has to deal with in respect to products is simply about item or SKU number, unit price, and delivery dates. These are all very specific discrete pieces of information that can be easily manipulated by a computer.

The other companies on our list also have elements of those characteristics. Automotive producers eventually have products that have vehicle identification numbers (VINs). General Electric has equipment and machinery that are identified by part and quantity. However, a great deal of these other companies' informational requirements are a lot more complex and revolve around product shape, descriptions, and definitions that, until recently, computers were not very well equipped to handle.

In any event, it is clear that the scale of companies is dramatically increasing, has increased over this period of time, and most likely will continue to increase into the future. PLM systems are not simply a luxury for these companies, but a requirement. If information systems are driving these companies to be able to substitute information for wasted time, energy, and material and to deal with increasing amounts of information accurately and reliably, then PLM is an integral part of these companies' requirements.

However, if we are not one of these rapidly scaling companies, this does not mean that we do not have to concern ourselves with this issue. The needs of these rapidly scaling companies will have derivative effects on the companies that they do business with. The larger these rapidly scaling companies get, the more likely that they will impact smaller companies simply because of the possibility that these smaller companies will have to do some of their business with one of these rapidly scaling organizations.

These rapidly scaling companies understand that the information requirements do not start and stop within their four walls, and so they are extending these informational requirements out through their supply chain. The automotive companies require their major suppliers to exchange and work with parts designs in electronic format. A request by an automotive supplier to an automotive company to, "courier over the drawings so we can quote on them," would only be met by peals of laughter and a sure loss of business by that supplier. Even where the informational exchange is simpler, say SKU and quantity, the major companies are dictating the technology. Wal-Mart is starting to require their suppliers to RFID[1] the products these suppliers send to Wal-Mart.

The change in scale means that, not only will the larger companies need to use PLM to deal with their product information, but PLM will trickle down through the supply chain and affect a much wider range of organizations. Even those organizations that will somehow be able to stay out of one of these supply chains will be affected. They will face the Hobson's choice of not investing in PLM and risking that the more efficient companies will eventually take their business or making the investment at the risk of their current profitability. Unfortunately, where change is the status quo, doing nothing is not a real option.

Complexity

The second aspect of change has to do with complexity. Not only are organizations getting bigger, but they are producing more and more complex products. To get a feel for that, we need only to look at Table 4.2, which is using the same time period: the 1970s to the 1990s. Table 4.2 represents the change in the complexity of product mix that today's organizations face.

What we are looking at here is the number of different products that are offered in the marketplace. The underlying premise is that, with an increase in the number of different products there is an increase in the amount of information about those products that organizations must deal with.

If we look at the 1970s compared to the 1990s, we see a number of diverse categories of products and the change in the number of products that are offered in the marketplace in those categories. Automobile manufacturers in the 1970s had 140 different models that they offered to the consumers. In the 1990s, they offered 260. In the related product category of sport utility vehicles (SUVs), there were eight different models in the 1970s. In the 1990s, this grew almost fivefold to 38 different models.

Table 4.2 Complexity

Product	1970s	1990s
Automobiles	140	260
SUVs	8	38
Running Shoes	5	285
Breakfast Cereals	160	340
Contact Lens	1	36

Source: P. David, "Understanding Digital Technology's Evolution and Path of Measured Productivity Growth: Present and Future in the Mirror of the Past," in E. Brynjolfsson and B. Kahn (Eds.), *Understanding the Digital Economy: Data, Tools, and Research*, pp. 49–98, Cambridge, MA: MIT Press, 2000.

But it is not only large ticket manufactured goods that show this increase in complexity. Running shoes went from having five models in the 1970s to having 285 models in the 1990s. This is a 57 times increase! Even such mundane products as breakfast cereals showed increases from the 1970s to the 1990s. Breakfast cereals increased from 160 different offerings in the 1970s to 340 offerings in the 1990s. New categories of products were not immune to the increase in number of the products they offered. Contact lenses increased from 1 model in the 1970s, to 36 different models in the 1990s.

We use the number of different product models as a proxy for complexity. Even if products are similar, the information and data about each product must remain distinct in its information base. While common components and platforms can serve to reduce some of that complexity, there is a great deal of work yet to be done, Industry lore is replete with tales of magnitude reduction possibilities: 40 fuel pumps where 5 will do, hundreds of fasteners where dozens will do.

There is also little doubt that the complexity of today's products themselves is increasing, although comparisons are harder to come by because of the fact that what used to be done by mechanical or electro-mechanical methods is done by programmable electronics. Electronics adds a whole new level of complexity to products. Compare the current Joint Strike Fighter (JSF) to the fighter jet of 30 years ago and the increase in complexity will be clear.

In addition, there has been another element of complexity added to the products of today. That added complexity is the appearance of software in today's products. The appearance of software has meant some decrease in the complexity of electronic circuitry because some functionality that was implemented with special purpose circuitry could be implemented with computer chips. It has also decreased some complexity in other systems, such as hydraulics, with the appearance of fly-by-wire and drive-by-wire.

However, the overall result of software's appearance has generally been to increase complexity as designers have taken advantage of software to dramatically increase the functionality of products. To convince ourselves of that, all we have to is compare the complexity and functionality of today's digital-based dashboards with the dashboards of cars from 30 years ago.

This complexity not only shows no signs of abating, but it is continuing to increase rapidly. The former Chief Technology Officer of General Motors, Tony Scott, recently predicted that while the cars of 1990 had approximately 1 millions lines of code in their systems, by 2010 that would grow to 100 million lines of code.[2] Unlike electronic circuits and components, software code can and does change frequently, which adds another major element of complexity.

The result of this complexity is the dramatic increase in the amount of data and information that needs to be created, cataloged, and monitored. We not only have to deal with a sheer increase in the number of different models of products, but with an increase in the complexity of the products themselves. When we add the additional requirement of tracking this data and information throughout the product's life, the size of the task becomes clear.

If companies were simply increasing in size by scaling a limited number of products, the increase in information would be a manageable problem. However, that is not the case. Not only are companies getting bigger by offering their products on an increasing scale, but they're offering significantly more products, which increases the informational characteristics of that organization by a substantial amount, possibly magnitudes.

While we haven't reached the mass customization of one, where every product is unique, what Table 4.2 is showing is that the number of products is increasing substantially, which also dramatically increases the requirements for information.

Cycle Times

The next element of fundamental change that is driving PLM is cycle times. Not only is the scale of the organizations getting larger and the number of different products or the complexity that they have to deal with increasing, but the timeframes in which they have to be completed are also shrinking dramatically. Long development cycle times for products are a thing of the past. Organizations are being forced by their competitors and their customers to create products on a much faster cycle time than they have in the past. For example, the automotive industry used to be on a design cycle whereby it had seven years from conception to production. Today, that design cycle is somewhere between two and three years, and the pressure is on to decrease that time frame. Automotive companies are now talking about an 18-month cycle, and even yearly cycles in terms of introducing new products.

But decreasing cycle times are not confined to durable goods with historically long lead times. In the clothing and fashion industry, the historical 90-day cycle times are no longer acceptable. This industry is working very hard to bring its cycle times down to fewer than 60 days.[3] The pharmaceutical industry has fundamental concerns not only about the minuscule success rate of its product development, but also about the lengthy cycle times needed to bring its products to market.

Decreasing cycle times have a dual impact. First, they substantially increase the amount of information that needs to be collected, processed, accessed, and stored. Even if decreased development cycle times mean decreased life-cycle times, which may not be true as products' durability improves, the information about the product needs to be accessible for an extended period of time because of rules and regulations regarding serviceability, regulatory, and/or litigation requirements.

Second, decreasing cycle time means the elimination of slack that previously existed to coordinate different uses by different functional areas, reconcile differences brought on by the use of inconsistent information, or keep the various functional areas' usage of information in synchronization. Decreasing cycle time means that the usual mechanism for dealing with disparate information—the coordination or review meeting where the project would stop while all parties checkpointed their progress and compared their individual progress to the overall progress—is a luxury in time that most organizations can no longer afford.

Instead, organizations must remain constantly in sync and not waste time by using inconsistent information that has to be reconciled and reworked. Organizations cannot afford the time to checkpoint and resynchronize their efforts. PLM is the new mechanism that facilitates this change in method due to these ever-decreasing cycle times.

Globalization

Globalization is also a major driver of PLM. There are two aspects to this globalization. The first is that globalization has affected the way we handle product information. The way we operate now is qualitatively different than in the past. The second is that globalization has provided us with new, sophisticated competitors from less developed countries which not only have cost advantages, but also access to the same technology the developed countries use.

In the not-too-recent past, it was possible to hold regular product development meetings where all the participants involved in the effort could huddle over the current product diagrams and specifications. The participants knew that that they were looking at the latest revisions, because all the revisers were in the one room. Products were less complex, so the chances of wasting resources by redeveloping products that were already developed or were being developed somewhere else in the organization were small.

Even large organizations like General Motors or IBM could convene these status meetings. In organizations this large, convening these meetings entailed some wasted people time since they might entail some local travel from satellite locations. The process was also somewhat inefficient in that parallel development needed to be paced around these regular status meetings to ensure that no area got out of synchronization with the project. However, cycle times and profit margins were sufficiently long and fat to accommodate this inefficiency.

But the days when the product information all resided in a single, physical area and everyone who needed to have access to it could walk over to a file cabinet or drawing vault and access it are long gone. In addition, companies now have design centers spread out not only across the nation, but across the globe. Twenty-four-hour design teams, where design activities follow the sun in a continual and constant fashion, are the goal—if not already in existence—for most major organizations. This "follow the sun" strategy has a design team that starts off in America and hands the design off to a Pacific Rim organization at the end of their workday. The Pacific Rim organization works through their day and then hands the design to their European counterpart at the end of their workday.

The European team picks up where the Pacific Rim organization left off and at day's end sends the latest update to their American counterpart. The cycle then starts anew.

This is the design ideal of the future, but elements are being incorporated into the processes and practices that companies operate with today. What this means is that product lifecycle management will need to be the repository of that information so that no matter where the individual is on the globe, he or she will be able to access and use that information—and more importantly, update it so that counterparts in the rest of the world can have access to it. With singularity of information, the term *hand-off* will become obsolete. There will be no actual hand-off of information as occurred in the past, but merely a hand-off of responsibilities in working with it.

The second aspect of globalization is that developed countries will now be faced with competition from less developed countries armed with capabilities and technologies that were previously a differentiator for the developed countries. In information technology, the general rule of thumb used in the 1970s and 1980s was that Europe was a few years behind the United States. The Pacific Rim was at least 5 to 10 years or more behind Europe.

This difference in capabilities was due to a number of factors. While the latest computer equipment was generally available to organizations in these areas of the world, organizations in the less developed parts of the globe often lacked the scale and resources to buy and deploy the expensive computers available. Even when that was not the case, the other elements necessary for a successful information technology deployment were often inferior or unavailable. These other elements included software, system analysts, programmers, an information-technology-savvy work force, and local training and education.

This is no longer the case. As noted in Chapter 1, the actual computer hardware has decreased in cost while increasing in capability. This means that scale is no longer the barrier it once was to information technology deployment. Increments in computing can now be bought by the inexpensive PC rather than the $1 million plus mainframe. In addition, the other necessary elements for successful information technology deployment are available.

Developing countries have not only educated and developed cadres of programmers and systems analysts who are available to local concerns, they now export that capability to leading U.S. firms in the form of outsourcing.

As a result, local firms in developing countries can acquire, deploy, and support the sophisticated capabilities of PLM. Organizations in developing countries are gaining the efficiencies that accompany the breaking down of barriers between functional areas and the sharing of product information in a comprehensive fashion. They are doing so with their lower labor cost advantage, although that may be compensated for somewhat by their disadvantage of scale.

Taken together, these two aspects of globalization require organizations to aggressively adopt PLM technologies and techniques. If they do not, they cannot implement a global approach to their workforce and risk falling irretrievably behind their global competitors.

Regulation

Government regulations are a different type of driver from the ones previously discussed. For the previous environmental drivers, organizations can choose whether or not to respond to them. The management of these organizations can make its own assessment as to the impact of these drivers and its response to them. It can choose to ignore these drivers or fail to respond to them. Government regulations are not in that category. Organizations must comply with government regulations or face immediate and severe consequences.

The increasing scope of governmental regulations requires organizations to deal with product information far differently than they have done in the past. In that past, when the product left the factory door, the manufacturer of products, to a great extent, stopped having to deal with them. While there were service organizations and warranty issues and things like that, the primary philosophy for manufacturers was that once the customer bought it, he or she owned it and was responsible for the product from that particular point on. *Caveat emptor*—let the buyer beware—was the overriding philosophy of the day.

However, over the last half-century there has been a change in how governments and their regulatory agencies have viewed their mandates. Governments have increased their regulatory power in areas of safety, external costs, and asymmetry in the balance of power between manufacturers and consumers

While safety was an obvious area of regulation, the other two areas were caused by a change in social perspectives. What were once external costs that manufacturers simply passed on to society (e.g., pollution, hazardous and toxic material use, disposal and landfill concerns) governments have decided should be an internal cost that manufacturers should be responsible for. Governments also have stepped into the relationships between manufacturers and consumers where there was felt to be an imbalance of power in favor of the manufacturers. For example, many states have passed so-called "Lemon Laws" that legislate the rights consumers have for redress in the event that the vehicles they purchase have a pattern of problems. This is irrespective of any contract the consumer might have signed.

This increase in regulation has forced manufacturers to dramatically revise their perspective of the scope of relevant product information. What was considered the responsibility of the product consumer yesterday is now the responsibility of the manufacturer. Manufacturers are being forced to look at the entire product lifecycle because regulatory agencies are insisting that they do so. If the definers of Product Lifecycle Management were overly ambitious in the scope of PLM, it really does not matter because regulatory progression is reaffirming that scope.

Examples of the new regulatory climate are as follows: the European Union Environmental Directives, the TREAD Act, EWRS, and the Sarbanes-Oxley legislation. This is not meant as an exhaustive list, but is representative of the scope and impact of this regulatory direction.

In terms of regulations and environmental regulations, the European Union (EU) is far ahead of the rest of the world with respect to requiring that manufacturers deal with their products throughout their lives. The European Union has issued the End of Life Vehicle (ELV) directive 2000/53/EC[4] that impacts all automotive companies selling vehicles in the European Union. What this regulation requires is that the total recovery of automotive vehicles be 85 percent of their weight by 2006 and, by 2015, 95 percent of the vehicle, by the vehicle's weight, will need to be recoverable. In addition, there is a requirement that there be no lead, mercury, cadmium, or hexavalent chromium after July 1, 2003. Furthermore, this directive requires that the product be designed for dismantling, reuse, and recovery.

This regulation forces the developer of the product, in this case automotive vehicles, to be worried about it, not only after it leaves the factory door or even the dealership's door, but long into the future when it needs to be disposed of and recycled. In this European Union directive, the requirement is that the automotive manufacturers set up certified recycling facilities where their products are not only recycled, but which also confirm that the recycling has taken place.

Automobile manufacturers will have to take responsibility for the recycleability of their products as they design them. What this also means is that it is no longer sufficient simply to define the bill of materials, but this will have to be extended further back to a bill of substances, which means that the individual material can be decomposed into its individual substances to see that it is meeting the requirement of these environmental directives.

While this is a European Union directive, because of the global nature of the automotive business, it is the concern of all the major automotive manufacturers. It will be very difficult for them to separate out their worry about the European Union versus their other design efforts for other parts of the world. It also seems that if these directives lead to a better environment, then other countries, specifically the United States, will follow suit because of the European Union's success. The argument to other regulatory jurisdictions about the costs of compliance will be difficult to sustain if they are complying with this requirement for European Union-based automobiles.

However, it is not only the automotive industry that the EU has targeted with its directives with respect to recyclability. The other major target is the electronics industry. The EU has also passed Directive 2002/96/EC[5] on Waste Electrical and Electronic Equipment (WEEE). This directive is similar to the ELV directive in that it requires manufacturers of electrical and electronic products to arrange for the recycling of their consumer products at the end of the products' lives at no cost to the consumer. In addition, this directive also prohibits the use of certain materials in electrical and electronic products. This requirement also means that manufacturers will need to know the substances in their products.

On November 1, 2000, the Transportation Recall Enhancement, Accountability, and Documentation (TREAD) Act was signed into law in the United States. The TREAD Act with its Early Warning Reporting System (EWRS) is a statute of the United States that calls for automotive companies to report safety concerns with their products on a proactive basis. It provides for civil and criminal sanctions if the automotive companies do not proactively search out information about the safety of their products and timely report it to the National Highway Transportation Safety Administration (NHTSA).

The third area of regulation is one that at first glance does not seem to be an area where product lifecycle management would come into play. This regulation is the Sarbanes-Oxley Act of 2002 or SOX. The Sarbanes-Oxley Act is a piece of legislation that resulted from the financial scandals of the late 1990s and is applicable to all public companies. Among other things, it requires that executives, under penalty of substantial criminal sanctions, certify the accuracy of their financial statements. However, that is not the only thing that Sarbanes-Oxley calls for. In addition, it calls for the same executives and their respective organizations to attest to the financial processes, controls, and systems that they have in place in their organizations to produce the financial reports. In addition, these financial processes, controls, and systems have to be audited by an outside audit firm.

As was stated in a *BusinessWeek* article written shortly after Sarbanes-Oxley was signed into law in 2002, "Starting next year, companies will have to prove that they can trace their product from assembly line to customer."[6] So, even though Sarbanes-Oxley is a financial act, it has substantial implications for organizations and their information technology systems. Product Lifecycle Management, which will assist in tracking these "as-built" products from assembly line to customer, will be an integral part of the information systems that executives will need to have in place in order to meet the statutory requirements of SOX.

An easy example of this issue is that most companies' work in process (WIP) is not in such a state that they can attest to the specific amount of product that is in process at any particular point in time. WIP is generally done on an aggregate basis. The amount of material produced is subtracted from the amount purchased and the beginning WIP balance. The computed result is the alleged current WIP amount. When periodic physical audits are performed, the difference between the computed amount and the actual count (and there is always a difference) is posted to a variance account.

As a result, what these companies report on their financial statements as work in process is a number that is based on a theoretical computation, as opposed to specific knowledge of what exists on the factory floor. Through a combination of enterprise systems, such as PLM and ERP systems, the requirement will be not to simply back into this number, but to have it derivable and verifiable at any particular point in time or state of the production process.

These are just some examples of the regulatory climate that is driving PLM. While it is difficult to predict the exact trajectory of regulatory increases, it is a pretty safe assumption that organizations will be required to track more information, not less information, especially where areas such as public safety are concerned. While governments appear to lag in their understanding of information technology, as evidenced by their own use of it, the requirements that they place on corporations show a keen understanding of what is technically feasible. As technology continues to evolve, we can safely assume that government regulatory requirements will keep pace. Organizations will have no choice but to comply.

Internal Drivers

Internal drivers are those drivers within the organization that are under the control or at least the influence of management. While the external drivers are out of the control of an organization's management, the internal drivers are something that management can focus on and impact. These internal drivers are areas that organizations must focus on if they want to meet the challenges of the external drivers. The main internal drivers affecting PLM are productivity, innovation, collaboration, and quality.

Productivity

As organizations use the term *productivity*, it takes on a qualitatively different meaning from the dictionary definition of "yielding results, benefits, or profits." As commonly used by organizations, "productivity" refers to the ratio of output that an organization obtains over the specific amount of inputs or resources it takes to produce that output. For tangible products, the output is in units, houses, airplanes, cars, computers, pills. The inputs are in human resources, denominated as the amount of time multiplied by a wage rate, and in material costs. As a simplified formula, Productivity = Units/(Time × Rate + Energy + Material + Information).[7] Since no capitalistic organization tries to decrease this ratio, discussions regarding productivity are always about the improvement of this ratio.

The quest for increased productivity is fairly universal. All organizations try to improve their usage of resources per unit of output. In some organizations that translates into improved profitability. They attempt to bring in a higher percentage of profits for the same amount of revenue. In doing so, these organizations will create more value for their shareholders, both in terms of short-term actual profits and longer-term enterprise value of the organization. Organizations are generally valued on a multiple of their profits, so increasing profitability has a multiple effect on the value of the organization.

In other organizations, the quest for profitability is not simply a strategy for creating value, but it is the difference between survival and demise. These organizations are under continual pressure to reduce the price they charge for their goods. They have to increase their productivity in order to decrease their costs in line with decreasing prices. The auto supplier industry is a prime example of this. Every year, automotive suppliers must deliver to the automobile manufacturers a 5 percent decrease in prices. If these automotive suppliers will not or cannot increase their productivity, then

bankruptcy is but a short step away as their prices decline—
paradoxically, sometimes even as their overall revenues increase.

The problem in increasing productivity is that the answer is
not in having people work longer or harder or buying material
cheaper. The low-lying fruit of productivity has already been
plucked. Most major organizations have embraced Six Sigma and
Lean Manufacturing in order address the inefficiencies in production and material purchasing and handling. In Figure 1.1, these
were what we called execution inefficiencies. To reiterate what we
said, execution inefficiencies, where we know what we need to do,
are what engineering excels at finding solutions for.

Outsourcing is another attempt to increase productivity. Outsourcing and its closely related derivatives, off-shoring and near-
shoring, increase the productivity ratio by decreasing the wage rate in
our equation above. Sometimes that can be accomplished by moving
the work outside the organization to a supplier with lower wage rates
because the supplier is non-union or in a lower wage area of the
country. In other cases, the work will be sent to a nearby country
(near-shoring) or a country farther away with a much lower wage rate
(off-shoring), with India and China receiving much recent attention.

There are a number of issues with this approach to productivity
improvement. The first is that the rate differential between countries can change over time. Canada's 20 percent cost advantage as a
near-shoring alternative has greatly diminished if not vanished in
2004 as Canada's dollar has strengthened against the U.S. dollar.
Facilities built in Mexico have been closed as production has moved
to China. Moving work in search of a lower wage rate can consume
resources that moderate the gain from a lower wage rate.

The second is that the complexity of coordination increases as
work is distributed to far-flung and constantly changing geographical
locations. That would also normally increase costs to counterbalance
the decrease in wage rate. However, the role of PLM is to provide
this coordination through the availability of product information.
PLM allows organizations to capture this source of increased productivity by moving the required information about the product to
wherever the work moves and to integrate the new, lower cost workforce with the use of the same processes and practices employed by
the previous workforce.

While PLM enables organizations to move around the globe in search of lower wage rates, PLM also enables a more sustainable productivity increase. This is the substitution of information for wasted time, energy, and material. Information allows the work force to eliminate wasted time searching for information, enables simulations to discover the most efficient processes, and facilitates reusing designs that normally would be unnecessarily duplicated. Information that replaces wasted time, energy, and material is a sustainable source of productivity that is driving the adoption of PLM.

Innovation

Innovation is another major internal driver of PLM. There are two distinct elements of innovation within most organizations. The first is product innovation. The second element is process innovation. Product innovation usually gets more of the attention, because the results of product innovation are the lifeblood of an organization—new products for the marketplace. Process innovation is generally a behind-the-scenes activity.

Where productivity is concerned with the cost side of the organization, product innovation addresses the revenue side. Without product innovation, an organization's revenue stream is at risk of declining, if not ceasing altogether. All too abundant are examples are of once-successful companies that fail to continue to innovate and are overtaken by competitors that introduce new and innovative products into their market space.

While innovation of product form is not to be underestimated—people seek and reward visual changes—the innovation of product function is the innovation that is the goal of most organizations. Novel products functions that create real value for their users by reducing the time, energy, and material required to perform tasks or by enabling their users to do tasks not previously possible are the innovations that organizations should focus on creating. These innovations are often confounded with function creep—adding functions that are novel, but do not add value because few people care about using them.

With cycle times decreasing, the pressure is on organizations to be more innovative in their product development. Quicker cycle times mean that new products have to be developed and introduced into the market. The competitor who has faster cycle times and get its innovative products to market quicker will take customers from organizations that have aging products. While cosmetic changes can constitute new versions of the product, relying on cosmetic changes is risky. The far better solution is to innovate.

Organizations are acutely aware of this. A number of industry studies show that innovation is a top, if not the top, priority initiative. Companies with a reputation for innovation like 3M Corporation have a stated goal that in a few years half of their revenue will come from products that do not exist today.

Innovation is due to the creativity of human endeavor and is not produced in a formulaic fashion. Innovation is also enhanced by the availability of the right information when it is needed. While innovation requires more than collecting, organizing, and coordinating product information from all phases of the product's lifecycle, PLM is an important enabler for successful product innovation.

Innovation also requires resources. Given that resources are constrained in any organization, one way that PLM contributes to innovation is by freeing up resources that might otherwise be wasted. If designers are wasting their time and the organization's resources by duplicating something that already exists, these are resources that, at least theoretically, could be spent on innovating new products.

Process innovation is closely related to productivity. The goal of process innovation is to find better technologies and methods in order to reduce the time, energy, and material that are required to produce the product. We will distinguish process innovation from productivity initiatives by focusing on the issue that process innovation initiatives develop alternate methods that use fewer resources. Productivity initiatives, on the other hand, simply identify and eliminate the wasted time, energy, and material in an already-known process.

While at one point in time, there was a belief that old proven products and the time-worn traditions of producing them were what made companies successful, we would be hard pressed to find an organization that either professed or practiced that philosophy. Companies understand that they need to innovate, but must innovate within the constraints of their resources. PLM enables them to be more efficient in their quest for innovation through better management of the information they have about products and free up resources that might be used in wasteful activities such as duplicating products that have already been designed once.

Collaboration

There is nothing new about the need to collaborate. Even to Og and Ug, our imaginary cavemen, collaboration was an instinctive and easily practiced trait when working together to drive a wooly mammoth over the cliff for dinner. Collaboration is not even a trait that we can call an exclusively human trait. Wolves do a very effective job in hunting as a pack to bring down a moose or elk many times bigger and more powerful than any individual wolf. This is clearly collaboration.

So why is there so much current discussion about and emphasis on collaboration, and why is collaboration a major driver of PLM? The simple reason is that we have changed the meaning of collaboration from working together at the same time and in the same place (which were the conditions for both our cavemen and the wolves), to working together over space and over time. This new definition of collaboration is neither instinctive nor easily practiced.

This was illustrated at a recent symposium organized by General Motors on the role of PLM in their organization. One of the slides they presented was a picture of their design center in the 1960s. It showed a large room with rows of drafting boards back to back. In the center of the picture was a group of engineers huddled over one of the drafting boards. They were collaborating.

The next slide that was shown at the symposium was a map of the world that identified all the current design centers across the globe. It did not show engineers huddled over a drafting board collaborating. It could not because, if collaboration was happening, it was happening very differently than it happened in the picture of the 1960s design center.

When we start to distribute work across the globe and engage in follow-the-sun design strategies, we need to concern ourselves with this new type of collaboration, because this is not the same collaboration that is instinctual and easily practiced. The farther we separate people in space and time, the less instinctual and easily practiced collaboration becomes. The farther apart in space and time that we separate, the more we lose the richness of communication and spontaneous cooperation that we take for granted when we are in the same physical space.

However, the increase in scale of our organizations and the move to globalization have disrupted our natural ability to collaborate. The premise that is behind the move to PLM with respect to collaboration is that if we cannot co-locate in physical space and time, then we should attempt to co-locate in our virtual space and time. PLM, with its singular view of data, its focus on creating a rich and complete view of the product through coherence of views, and its contemporaneous reflection of changes to the physical product in virtual space, attempts to re-create in virtual space the richness of communication that collaboration requires.

Quality

The lack of quality is another way of describing wasted time, energy, and material. There are two aspects to quality. First, quality is the characteristic of the product meeting its specifications. The second aspect of quality is performing to a particular standard of usage. The first aspect is controllable. The second is often not.

Over the past 30 years, organizations have focused a great deal of attention and energy on quality. A good reason to do so was simple economics. Products that met their specifications did not need to be scrapped, reworked, or repaired before they could be sold. Organizations also began to demand this same level of quality from its suppliers because, even if the manufacturer rejected the shipment from a deficient supplier and refused to pay for it, the costs of receiving, inspection, storing, and rejection were unnecessary costs that negatively impacted overall productivity. As manufacturers have moved to Just-in-Time (JIT) manufacturing, the ability to tolerate a substandard shipment no longer exists because the slack once provided by inventory stocks is gone.

In addition, organizations have increased their exposure to quality issues by competing on the issue of quality. How do they convince their product buyers that they have a better quality product?

They do so by warranting the product for a longer period of time. By doing so, organizations pick up the costs of insufficient quality that were once the costs of the product purchasers. If the quality is good, then there is no additional cost. However if the quality is not good, then there will be added costs.

One cause of quality issues is a lack of information as to what the specifications are. With different version of the parts or an inability to fit different parts together, the opportunity for inconsistent or inaccurate specifications is prevalent. PLM, with its consistent and singular view of the product and its components enables all the producers to know and understand the specifications. In addition, PLM's ability to convey the specifications visually takes the ambiguity out of product specifications.

The second aspect of quality, and probably the more important of the two—the ability of a product to meet a certain standard of usage—is much less controllable and more problematic. Products that meet the specifications but do not meet the usage requirements will make it successfully out the factory door, only to return. This will add additional, but wasted, cost.

As we will discuss in Chapter 6, the problem with this issue of quality is that it is difficult to perfectly map requirements to specifications, especially where the requirement is a clear but imperfectly specified requirement such as "must not kill or maim owner when used under normal conditions." The difficulty is that, while all designers know what this requirement means, it is hard to know how they can map it onto a set of specifications.

The problem is that with complex products operating in complex environments, all the possibilities of operational states cannot be known by analysis. Instead, extensive and exhaustive testing of different product states under different conditions needs to be done. Done in real space, this is a costly and time-consuming endeavor. PLM can assist by using its different virtual spaces to test a wide range of product states and conditions in substantially less time and at lower cost. In addition, it can test many more possibilities than can be done in real space. Again, substituting bits for atoms eliminates wasted time, energy, and material.

Boardroom Driver—IT Value Map

While all these drivers of PLM are interesting, they must be translated into economic terms in order for organizations to evaluate and act upon them. Because of its impact on so many different functions, the decision to invest in PLM logically is made at the upper levels of the organization on the basis of the quantified value it brings the organization. It is interesting to understand the drivers that are moving organizations, but, in the end analysis, if the value PLM brings cannot be quantified and explained to the executives and board members who approve capital expenditures, it will not be adopted.

Although there are qualitative measurements of value, such as the reputation of an organization, the quality of work life for its workers, and its standing in the community, the decision to invest in PLM or any major information technology initiative of this sort is made on the basis of a financial analysis. The most common measurement of this is Return on Investment (ROI), which is simply the increase in income of an organization as a result of the initiative, divided by the amount of resources that the organization needs to invest in that initiative.

The issue with ROI is that it often includes the allocation of resources, such as personnel, space, overhead, etc., that already exist within the organization. The effect is that there is really no new investment being made. The costs incurred by the organization remain the same in the short run whether or not the project is undertaken, although in the long run those resources may be eliminated if initiatives and/or projects are not undertaken to use them.

For that reason, we will use Return on Assets (ROA), which is the change in income divided by the cost of the asset, as our comparison benchmark. Adding an asset usually requires expending cash, which is a carefully watched resource in even the biggest, most successful organizations. A major expenditure of cash requires the approval of the senior executives and/or board of directors, whereas reallocating existing resources usually does not. This approval group will be looking for an impact in income in order to justify adding an asset to the balance sheet. We will use ROA, the increase in income resulting from the deployment of the asset, divided by the investment in that asset, as a more stringent measurement than ROI.[8]

Unless one is trained in accounting and enjoys it, the profit and loss statement from which this change in income is derived can be daunting both to look at and to understand. In addition, the senior management and directors who evaluate and approve these investments have neither the time nor the inclination to immerse themselves in the accounting details. They look for simple methods that display the larger picture and allow them to understand the linkages that drive changes in income. To that purpose, the author has developed what is called the IT Value Map that gives, in a simple yet powerful graphical manner, the perspective and impact of various factors that result in changes to income.

Income, Revenue, and Costs

Figure 4.1 presents the basic IT Value Map. If we look at the left-hand side, we see that the final outcome is value. For initiatives to get the attention of senior management and the board of directors, initiatives need to show that they can create value for the organization. As mentioned before, while there may be qualitative measures of value, such as market share, quality of work life, perception in the marketplace, etc., senior management and their board of directors invariably make their investment decisions regarding initiatives on the projected financial impact. In fact, for public companies, the

Figure 4.1   IT Value Map

board of directors has a fiduciary responsibility to its shareholders to place their interests above those of other stakeholders.

This means that this value is cast in terms of the change or improvement in income over the cost of the assets that will be needed to obtain this improvement of income. The decision to approve a project is based on there being an increase in income that not only offsets the cost of the new assets being acquired, but also generates a new earnings stream that justifies investing in that particular asset rather than some other asset. In all organizations, resources are limited and organizations try to allocate their capital expenditures in rank order from the highest returns to the lowest ones.

As shown in the IT Value Map, there are only two ways to increase income. The first is to increase revenue. The second is to decrease cost. If we look at the revenue side, which is in the top area of the IT Value Map, revenue basically consists of two components. One component is the price of the product that the organization is producing. The other component is the quantity of the products sold.

All things being equal, the price of that product is dependent on the quality of that product and the amount of functionality that the product has. The more functionality a product has, the higher its price. The better the quality of the product is, the higher its price. However, it is important to note that quality can also be negative. A company with declining quality can see the price for its product decrease.

On the cost side, at the end of the day, there are really only two things that organizations buy. They buy people in the form of their time and rate and they buy material. While organizations are divided into functions and departments, the end result is that, even in the most complex cases, organizations are simply buying two things: (1) people's time at a specified rate, and (2) material at a specified cost. In fact, even when organizations are buying material, they are rarely purchasing just material alone. Included in the purchase price of the material is the time of the supplier's people multiplied by a wage rate, plus—hopefully for that supplier's shareholders—a profit. This relationship continues down the supply chain.

Direct costs usually come with a multiplier. Overhead percentages are usually relatively stable with respect to direct costs. When we add direct costs, we also eventually add a percentage of indirect costs for every dollar of direct costs we add. When we add people, we add costs in the form of their time multiplied by their rate. However, we also add other costs such as more supplies, more space, and more material. We also add other people at their time multiplied by their rate. These people process the new payroll, sell the additional products produced by our new people, bill for the new revenue, manage the new people, etc. These people need supplies, space, and material, and so it goes.

This relationship is natural on the upside. People are vocal about pointing out the need for more resources when costs are added. However, people are not so vocal in pointing out that excess resources exist when costs decrease. Recovering the additional overhead costs must be demanded when costs decrease. Inertia always sets in, and decreasing the direct costs of people's time rarely delivers an automatic corresponding multiplier. When looking at the IT Value Map, this multiplier needs to be factored in.

If we expect to increase the value to the organization, we would expect our investment to decrease the time and rate of people and the cost of material. If we are looking to the revenue side for our increase in income, we would expect our investment will result in the increase in the price and/or the quantity of our products that our customers are buying. Either a cost decrease or a revenue increase improves income and should allow us to justify our investment in new assets.

That is in theory. The reality is that it is difficult to justify an IT investment on the premise that an increase in revenue will increase income. We do not have the control over revenue that we have over costs. While we can control costs, we can only project income. We can cut headcount. We can freeze hiring. We can limit purchase orders. But we cannot require our customers to pay a higher price or buy a greater quantity.

The issue is that IT is an enabler of increased revenue, but that it takes other action to realize it. IT initiatives can only provide the information. It has to be translated into new functionality or better quality. So while there is a hope, a promise, even an expectation of increased revenue, IT initiatives usually must be justified on their ability to deliver income increases through cost savings on the current revenue base.[9]

Before we look at some examples of the use of the IT Value Map, it is important to put its use into perspective. While the IT Value Map is a high-level representation of the effect of IT initiatives, it does not replace the detailed financial analysis that is always required. The IT Value Map is meant to illustrate the results of that analysis. However, it will identify whether costs are simply being moved around because there will be no net savings of material or people's time and/or rate. Also, any of the cost areas can be broken down on the IT Value Map to isolate the specific areas where cost savings should be realized.

Comparing Lean Manufacturing, ERP, CRM, and PLM

Now let us look at some actual examples analyzing the impact of IT initiatives using the IT Value Map. Figure 4.2 is the IT Value Map for Lean Manufacturing. Lean Manufacturing is generally a successful project for organizations. The reason for this is that, on the asset side, there is not a significant investment in assets. It really just provides training for people on how to employ Lean Manufacturing techniques and arms them with an IT infrastructure that can measure costs accurately. The increase in income derived from initiating a Lean Manufacturing project comes from a decrease in

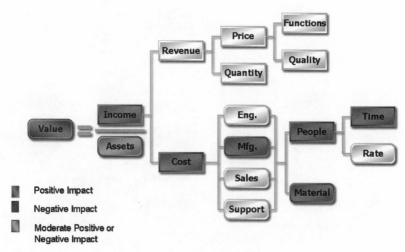

Figure 4.2 IT Value Map—Lean Manufacturing

costs. Those costs are limited primarily to the manufacturing department, although the philosophy is being exported to other areas: Lean Office, Lean Engineering, and Lean Development.

What Lean Manufacturing does is to decrease the amount of time that people waste on inefficient tasks. In addition, Lean Manufacturing attempts to decrease the amount of wasted material due to the production of incorrect or excess product. As a result, Lean Manufacturing generally has a very good value proposition for the organizations that take on Lean Manufacturing projects because it can decrease wasted time and material. This therefore decreases costs, with a resulting increase in income.

If we look at Enterprise Resource Planning (ERP) initiatives in Figure 4.3, we can see that the situation is somewhat similar. Although, in this case, ERP also proposes that it has an impact on the amount of revenue that an organization can get because it improves the quality of the product. All things being equal, increasing the quality of the product should improve the price and, therefore, improve revenue

On the cost side, ERP is primarily a manufacturing and sales system, and its impact, not unlike Lean Manufacturing, is to decrease the amount of time and wasted material that people spend

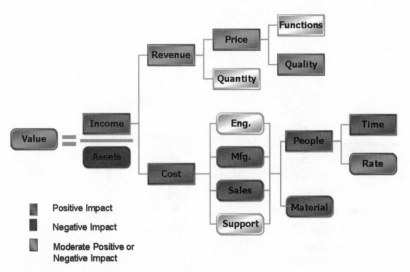

Figure 4.3 IT Value Map—ERP

on the project. The mitigating factor for ERP is collecting information. It's not free, as we've seen above, and therefore there might be an increase in the rate because the sophistication of the people employed in the data collection process will generally increase.

However, ERP systems are generally very expensive. Therefore, the increase in income through the decrease in cost and increase in revenue is offset by the substantial increase in assets that are deployed to implement an ERP system. If the increase in income is not sufficient to offset the substantial increase in cost, ERP systems can turn out not to be as successful as their proponents would like. ERP systems are primarily a cost-decreasing system and the departments they affect are the manufacturing and sales departments.

Customer Resource Management, or CRM, systems have had a spotty reputation for success, and the reason can be seen in the IT Value Map in Figure 4.4. Basically, CRM proposes a race between revenue and cost and, if the race is not won by revenue, then CRM systems cannot increase income. In fact, they may result in a decrease in income. As we can see, the revenue is impacted by the customers and the prices of the products that they buy. What CRM systems attempt to do is impact the price under the assumption that the quality of the product should improve because organizations have a better understanding of the products that their customers want. But the real value of CRM is increasing the quantity of product sold, both through increasing sales to existing customers and to new customers by better understanding of and ability to meet their needs and requirements.

On the cost side, CRM systems are primarily sales systems. But because the information that they're collecting is not free, the impact of CRM systems is to increase the amount of time people spend on collecting this information. And because CRM systems are not uncomplicated, generally the sophistication of people, and therefore their rate of pay, will increase. We can see that the cost aspect for CRM systems is negative. Thus, this is a race between revenue and cost increases. Even if revenue wins the race, the resulting net increase in income needs to be weighed against a not insubstantial cost of the software investment.

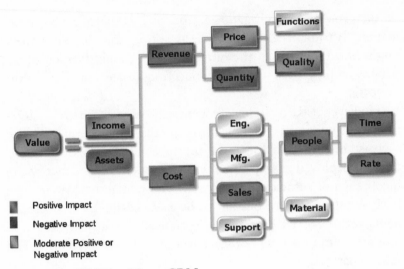

Figure 4.4   IT Value Map—CRM

If we look at Figure 4.5, we can see the impact that PLM has on the organization. We can see at first glance that, unlike the other initiatives we have looked at, PLM impacts all aspects of the organization. We expect PLM to have a major impact on income. However, unlike some of the other IT initiatives, PLM is not simply a matter of trying to decrease costs, PLM also has a focus on increasing revenues.

With respect to the lower half of the IT Value Map concerning costs, PLM has an impact on all the functional areas: engineering, manufacturing, sales, and support. PLM is focused on creating Lean Thinking across the entire organization by substituting information for wasted time, energy, and material. This information enables people to reduce the time spent on developing products that already exist. People can spend this time working on improved versions of the products that minimize manufacturing costs and are easily serviced. This information will allow companies to reduce the material used in making duplicate products and producing products that are known to have defects discovered in the field and that have since been corrected in the design specifications, but are still being rapidly produced on the plant floor,

The increase in income is also impacted by the top half of the figure that deals with the revenue impact. Two of our strategic

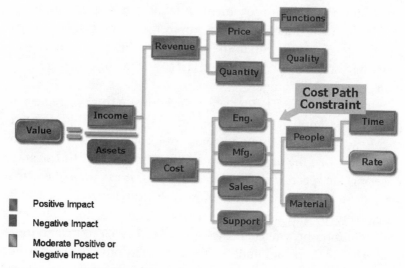

Figure 4.5 IT Value Map—PLM

drivers, innovation and quality, are at work here. With the approach that PLM provides, we would expect that the price the firms can charge for their products should increase. This is because both the quality and the amount of functionality that organizations can build into their products, given the same amount of resources, should improve.

With respect to increased functionality, if organizations do not have to keep recreating the same or equivalent components or parts of their products, they can focus their attention on improving the functionality of the products with those same resources. With better information, organizations can produce better-quality products.

By using virtual space to better design, validate, and test their products, organizations can have a higher confidence level that their products will perform for their users the way they are supposed to. Products that perform as well or better than users expect are the real definition and test of quality. Increasing market share results in an increase in the quantity of units sold and is a natural consequence of better quality and functionality.

As the left side of the IT Value Map illustrates, the creation of value by PLM is not only a function of the increase in income through reduced costs and improved revenues. It is dependent on the relation of this improvement in income to the investment in PLM. The investment in PLM is not insignificant. For large organizations it can easily run in the millions of dollars to acquire software, hardware, and consulting services and change management, education, and training.

However, unlike ERP, which requires the up front commitment of millions of dollars worth of resources to support this initiative, PLM can be phased in on a project-by-project basis. This means that even when the organization develops an overall PLM strategy covering the spectrum of PLM capabilities, it can implement that strategy in distinct initiatives, ranking the initiatives on their individual returns and making the necessary investments over time. PLM solution providers, as they attempt to reach more and more organizations, are unbundling their applications so that smaller organizations can acquire and implement affordable subsets of PLM. For smaller organizations, investments in PLM can now start at tens of thousands of dollars.

As noted before, it is difficult to justify IT initiatives on the basis of revenue increases. However, based on the widespread impact of Product Lifecycle Management initiatives within the organization, the chances are good that organizations can find sufficient cost decreases by using PLM systems and justify a sufficient ROA on cost savings alone. Even though revenue increases alone will usually be insufficient to justify PLM initiatives, the promise of revenue increases will provide an incentive to approve even PLM initiatives that have marginal cost-based returns.

Finally, as noted in Chapter 2, the estimate is that as much as 80 percent of the product's costs are set at the point in time that the product is designed and engineered. Thus, an initial focus of PLM in the engineering department is logical and highly warranted. It is appropriate to apply the PLM approach to the designing and engineering of products and to attempt to affect the cost structure of these products at their early stages.

From looking at the IT Value Map, it is apparent that PLM is a corporate initiative because it affects all areas of the organization. While PLM-related initiatives can be beneficial within functional areas, it is important not to lose sight of the fact that the real payoff is when PLM initiatives cross functional boundaries. In addition, while PLM is certainly about thinking lean by decreasing the costs of wasted time, energy, and material, PLM can also have a major impact on the revenue side of the equation. PLM enables increasing complexity, decreasing cycle times, the fostering of globalization, and regulatory compliance. However, PLM will also enable companies to continue to increase their scale by improving the focus on innovation, collaboration, and quality. As shown in the IT Value Map, PLM is the initiative that can have an impact on all aspects of the organization.

Summary

In this chapter, we discussed the environment that is driving PLM. On the external front, we have the increasing scale of organizations, the increasing complexity of products, the inexorable march toward decreasing cycle times, the role of globalization, and a driver that cannot be ignored—regulatory change.

On the internal front, we see organizations wrestling with a need for continuing productivity for many just to survive. Organizations also have a focus on innovation, collaboration, and quality as means to grow and thrive.

Finally, we examined the primary driver of interest in PLM for all organizations: its impact on the financial results. We introduced the IT Value Map as a method to understand at a conceptual level the value of PLM, and compared the impact of PLM with other IT initiatives. From that analysis, it is clear that PLM is an initiative that impacts the entire organization and impacts both revenue and costs. In addition, the initial focus on PLM at the first stages of product development makes logical sense—but this should be only a first step. The real value of PLM comes from its cross-functional application throughout the entire organization.

Notes

1. RFID stands for Radio Frequency Identification. These are electronic devices that contain information about the object they are attached to. Unlike bar codes, which they are intended to replace, they can be read at a distance. For more information, see S.E Sarma, S.A.Weis, and D.W. Engels, *RFID Systems and Security and Privacy Implications*, Cambridge, MA: Auto-ID Center, MIT, 2002.

2. See R. McMillan, "GM CTO Sees More Code on Future Cars," *InfoWorld*.

3. See G. Kahn, "Making Labels for Less," *Wall Street Journal*, August 13, 2004, p. B1.

4. For the actual European Union End of Life Vehicle directive refer to http://europa.eu.int/eur-lex/lex/LexUriServ/site/en/consleg/2000/L/02000L0053-20030101-en.pdf. Accessed 07/09/05.

5. For the actual European Union directive refer to http://europa.eu.int/eur-lex/lex/LexUriServ/site/en/consleg/2002/L/02002L0096-20031231-en.pdf. Accessed 07/09/05.

6. See D. Henry and A. Borrus, "Honesty Is a Pricey Policy," *BusinessWeek*, October 27, 2003, pp. 100-101.

7. This is of course the cost function from Chapter 1. For purposes of this discussion, energy, material, and information are assumed to have been costed so are denominated in dollars. We are explicitly costing people's time as their time multiplied by their rate. While up to now, we have treated material and energy as being separate, we will use "material" as referring to both energy and material. It will simplify the diagrams later in this chapter without any loss of distinction. Energy is often purchased as material, e.g. coal, oil, gasoline, and one of the greatest minds of all times has established the interchangeability between energy and material, Einstein's famous $E=MC^2$.

8. Ideally we should use a cash-to-cash comparison, with the increase in cash resulting from the investment in the asset divided by the cash expended to acquire the asset. However, organizations use accrual accounting, which hides the cash effect. This is unfortunate because as is said, not incorrectly, "Cash is reality. Income is merely some accountant's opinion."

9. The major exception to this rule was during the dot com era when all sorts of projects were justified on solely on the basis of new and increased projected sources of revenue. There was no consideration given to cost savings. In fact, these initiatives were not considered serious unless accompanied by major new spending. The term, "burn rate," a term naturally used a few miles north of me at Cape Canaveral came to be unnaturally used in the board room to describe the spend rate while waiting for the new revenue sources to arrive – many which never did. That era has ended. See J. Cassidy, *Dot.com: The Greatest Story Ever Sold* (1st ed.), New York: HarperCollins, 2002, for a good description of that brief but interesting era of IT.

PLM Elements

W E CAN'T SIMPLY buy PLM as a software application, although software is an integral component. This chapter will describe how people, technology, and processes/practices are all integral to a perspective of PLM. It will explore the characteristics of each of these and explain how they fit into a PLM framework. Chapter 5 will also focus on the issue of practices, which is often ignored in favor of a focus on processes.

When we reviewed definitions of PLM in Chapter 2, we saw the terms *approach* and *system* used. What are the elements that make up this approach or system? We propose that PLM is made up of people, processes/practices, and technology.[1]

The Process/Practice versus Technology Matrix

When looking at PLM, it would be easy to look at the technology first. After all, information technology is the new element in the mix of people, process/practice, and technology. While people and their processes and practices have been around a long time, information technology is relatively recent and novel. In addition, information technology is a key enabler in that it allows us to do tasks that are not impossible, but surely are impractical, without this technology.

However, with respect to PLM, a focus on the information technology without first considering people and their processes and practices is ill considered. While information technology may be a key enabler, people and their processes and practices are a prime necessity in implementing Product Lifecycle Management. In PLM, technology does not completely replace people and their processes and practices as it might in other uses of technology, such as robotic manufacturing or automated insertion and assembly of electronic circuit boards. Instead, information technology enables people to perform their processes and practices in a much more efficient manner.

To get a better feel for this relationship, we can look at Figure 5.1, which shows a matrix of the impact of low and high information technology utilization against low and high process/practice development. By *information technology*, we are referring to all manner of information technology, from manual paper cards and records to sophisticated computer and software applications and systems.[2] *Low information technology* refers to the lower end of this continuum, such as manual paper forms and records, while *high information technology* refers to online computer-based systems. *Low process/practice development* refers to processes and practices that are ad hoc, undefined, or spontaneous. *High process/practice development* refers to process/practice development that is continually being analyzed, refined, and improved.

In the lower left quadrant, we have low information technology–low process/practice development. Unsurprisingly, the result

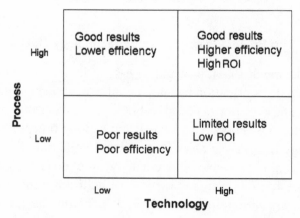

Figure 5.1 PLM Project Outcomes

of this combination is an organization that has a great deal of wasted time, energy, and material. It exhibits very poor efficiency, especially where there is high employee turnover. Like our quasi-amnesiac tribe, each day is a brand new day. Even what should be routine, standardized tasks are performed differently by different people, and a great deal of time they are performed differently by the same people. In situations where there is high turnover of people, there is a great deal of trial and error, with error winning more times than not. Any information learned is immediately lost, with the errors destined to be repeated over and over again. The waste of time, energy, and material is high.

If we shift one quadrant to the right, which is the lower right quadrant, we have the situation of low process/practice development and high information technology. This is not an uncommon situation. Many organizations, when faced with the situation of low process/practice development coupled with high employee turnover, look to information technology as a panacea. Usually, they are disappointed.

What they wind up with is highly automated poor processes and practices. While the technology may force some discipline and consistency, automating poor processes and practices generally does little to improve the efficiency of an organization. In some cases, the new system will be treated as a nuisance that is required by the job.

If this is the case, employees will exert the least possible amount of effort in working with the system. They will treat it as an unnecessary burden. Entering data will be the task they most often defer. They will batch information that should be entered continuously. When they enter the information, they will enter the least possible data, entering any single character to circumvent required checking of entry fields and abbreviating everything wherever and whenever they can. They will have no interest in or use for the information developed by the system. They view the information system as something to be beaten.

In rare cases, there will be some useful structuring and discipline imposed on the organization in this quadrant through the use of the information technology system. However, because the process and practices are so poor and ill-defined, the obstacles, bottlenecks, gate keeping, and delays will still be present, so that any task-structuring the information system brings will be lost in the overall inefficient process and practice.

The result is that this low process/practice development and high technology quadrant is characterized by limited results and low return-on-investment (ROI). The low return on investment is a result of the efficiency gains from task improvement being dissipated by the poor processes and practices that the technology is implementing. Especially in approaches such as PLM where the information requires contextual understanding and judgmental decisions, implementing technology where poor processes and practices exist is sure to lead to underwhelming, if not counterproductive, results.

If we move diagonally left and up, we move into the high process/practice development and low technology quadrant. This is an interesting quadrant because this combination can produce good results. With well-defined processes that are continually improved and low information technology, organizations can and do get good results. Good results mean a continual increase in efficiency and a decrease in the waste of time, energy, and material.

The best example of this is Toyota's Total Production System (TPS). As part of TPS, Toyota uses simple information technology such as kanban cards that pull inventory from the previous cell. While the technology is simple, Toyota relentlessly continues to analyze, improve, and document its processes. Toyota also spends a great deal of time and effort to train its people on the processes.[3]

While some might think that this approach would be successful in a deterministic setting where processes ruled—such as a factory—but would not be successful in a setting where practices were predominate—such as the design area—Toyota has proven them wrong. Toyota has also been successful in eliminating wasted time, energy, and material in the design and engineering function. Toyota has shaved years from the design and engineering cycle, driving the design and engineering cycle to less than two years.

However, Toyota appears to be a very unique company that is difficult to emulate. Toyota also realizes that the low technology approach means that it must use more people than it would if it used information technology. It understands that experimenting with processes and practices in real space—moving atoms—as opposed to experimenting in virtual space—moving bits—is not as efficient. Toyota has made and continues to make substantial investments in PLM information technology.

The reason that Toyota does so is the last quadrant. The final quadrant, high process/practice development and high technology, is what organizations strive to reach. In this quadrant, they are using information technology to implement continually improving processes and practices. With respect to the information-processing component, technology is more efficient and reliable than a solution implemented using only people and paper. In addition, processes and practices can be simulated in virtual space, saving the cost of wasted time, energy, and material in moving these equivalent atoms in real space.

The People, Process/Practice, Information Technology Triangle

The process/practice versus information technology matrix is conditioned on one main factor: people. The condition under which this matrix holds true is that people are motivated and act with good intentions and competence in making the process/practice and information technology work. If they do, then this matrix holds true. If they do not, then the best processes, practices, and information technology will fail. Willful people with ill intent defeat processes, practices, and information technology every time. Well-intentioned, motivated people will make mediocre and even poor processes, practices, and information technology work for them—although, in that case, expect a fair amount of improvising!

Figure 5.2 shows the elements of PLM. To illustrate our point of people dominance graphically, this figure is a triangle that has a base of processes/practices and information technology. People are at the top of the triangle to indicate their dominant role. However, all three elements must perform in their own right and work together cohesively to make for an effective implementation of PLM. We will discuss each of these elements of PLM and their characteristics below.

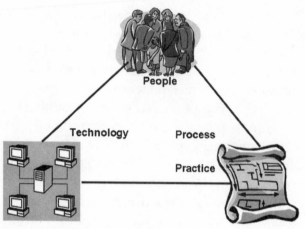

Figure 5.2   Elements of PLM

People

With respect to PLM, what characteristics of people do we need to consider in order for PLM to be successful for an organization? What characteristics will help them in this endeavor and what characteristics will limit their effectiveness? We believe that the important and relevant characteristics of people that we need to concern ourselves with are: capabilities, capacity, and organization. As an approach, PLM has to take these characteristics into account when assessing whether PLM can succeed or fail within an organization.

Individuals within an organization come equipped with a wide range of capabilities. Some have limited capabilities suitable for simple, uncomplicated, well-defined tasks. Others have very robust capabilities that allow them to engage in efficient goal achievement even under ambiguous and uncertain circumstances. Irrespective of where they fall on the spectrum of capabilities from limited to robust, the capabilities of people are for the most part determined by these aspects: experience, education, training, and support.[4]

Experience

Experience is accumulating information and knowledge about different situations in order to be able to trade off information for wasted time, energy, and material. As human beings, we instinctively do this every time we experience and learn from encountering a situation or performing a task. If we are ambitious, we figure out quicker and easier ways to do the task as the time progresses.

With respect to PLM, our experience allows us to reduce search times, predict outcomes with a higher probability of success, and pick the right routines for the right situations. If we have experience in the categorization of parts, we can look through only those categories of parts that perform the function we want to add to a new product. We have experience in knowing what tool paths will machine the part feature we want. We have experience with the order of assembly so we know which of the sequences can be done—and be done in the least amount of time.

Experience is an individual characteristic. If the individual leaves, so does the experience. However, if we can capture the experience by having the experiences recorded in virtual space, we have the opportunity to impart that experience to others. While we have long done this through books and other educational activities, PLM takes it a step farther by embedding experience into processes and routines. As we will see in the next chapter, "start" or "smart" parts are an outgrowth of this effort to capture experience.

However, we have to be a bit careful about assessing experience. Oftentimes people with 30 years' experience really have 1 year's experience for 30 years, and as a result the experience level they have is equal to the person who has been on the job for 1 year. If the individual with 30 years' experience has been performing uncomplicated, standardized tasks with little variability, he or she might reach the limit of learning in a relatively short period of time. The additional time on the job adds little or no information or knowledge and may ossify attitudes as it continually reinforces a pattern of behavior. As a result, we have to assess the variety and range of experiences and not simply equate duration with experience.

With respect to PLM, experience is a two-edged sword. Because we can trade information for time, energy, and material, our most experienced people will normally have the most information about our products. Capturing that information as they work with the PLM applications will improve the efficiency of the transition to PLM.

A number of organizations are wrestling with a common problem related to this. They have a knowledge gap in their workforce.

They have a large group of baby boomers retiring in the next few years. They also have a contingent of inexperienced, younger workers hired in recent years. What they do not have are workers in the middle because of the downsizing phenomena of the 1980s and 1990s. If these organizations hope to capture the experience of the soon-to-be retired workers, their only hope is PLM.

However, PLM also requires new and different ways of approaching product information. Our most experienced people may also be our most rigid people, who may resist new ways of doing things because they have "always done it the old way." PLM's promise is not in duplicating the old ways, even if it utilizes the cheaper bits versus the more expensive atoms. PLM's promise is in synthesizing a new way of doing things. However, this will require flexibility on the part of people. Flexibility is not always a trait compatible with experience.

In addition, organizations no longer have the luxury of the time and cost of waiting around for people to gain experience. Increasing product complexity and decreasing cycle times means that we need to gain more experience in less time. In addition, gaining experience is an expensive, passive, unstructured activity. It is expensive because we may have to spend a good deal of time, energy, and material to find the right solution for a problem through trial and error. It is passive, because activities where we gain experience happen when they happen. It is unstructured, because we do not plan it. Experiences happen when we least expect them, and a good deal of the time when we least want them.

Education and Training

A way to get around waiting for people to gain experience is to provide them with education and training. Education and training are classical applications of trading information for wasted time, energy, and material. We make the distinction between education and training in the following way. With training we teach people what to do. With education we teach them why they do it.

Training is better suited to processes. We want the same actions performed each and every time so that we get the same results each and every time. Education is better suited to practices. For practices, we need to understand the theories of inputs affecting outputs so that we can separate the relevant factors from the irrelevant ones. This will allow us to understand complicated situations and create novel and innovative instances.

The information that we collect within PLM can greatly enhance the effectiveness of education and training. Using the virtual spaces of the Information Mirroring Model, we can develop proactive and structured simulations that will have the same effect as experience for the individual at substantially less expense. These simulations can utilize actual situations and be more than theoretical exercises.

Simulated activities have always played a role in training for processes, but they can also be useful in education for practices. An example of this in PLM activities is the categorization of designs so that they can be retrieved and used by other engineers. New engineers need to understand the taxonomy of the type of designs that are used within the organization and the different search parameters that are useful in retrieving those drawings. We can educate the engineers on those and then let them try to categorize drawings on their own. We can then let them see how their categorizations compare with the categorizations that experienced engineers have done with the same designs. They can continue this simulation until the categorization of the new engineers matches the categorization of the experienced engineers.

Support

Finally, the support that an individual is provided with will also enhance or detract from his or her capabilities. Support is an extension of education and training, but takes place during the execution of the task, not in preparation of the task. Support is also a substitution of information for wasted time, energy, and material. Unlike computers, which once programmed can retrieve the necessary information at any time, people tend to have diminished recall of information the longer they have not used it.

The proper support can reduce the inefficiency of the searching and relearning process by providing people with the information they need when they need it. If processes are only used periodically (such as year end), or infrequently (such as a request for a manufacturing process variance), we should provide support to avoid wasted time in searching and relearning. Because the applications supporting PLM are complex and affect complex processes, context-based support is required. If it is not provided, people will become inefficient and frustrated with PLM when required to perform tasks that are done infrequently.

For practices, it may not be enough to provide support through computer applications alone. We should use these applications to provide rich communications between individuals. In a book by John Seeley Brown, the former director of Xerox PARC, he presents an interesting twist on an old adage. In investigating how copier repair technicians solved thorny problems, he concluded that when the going gets tough, the tough go to coffee.[5]

This was his way of encapsulating observations on the practice of technicians solving problems where the inputs or symptoms were ambiguous and not well defined. Although it might have been taken as a time-wasting social event, morning coffee was really a practice where these technicians collectively diagnosed and developed solutions. Eliminating these coffee breaks in the name of efficient processes would have produced the opposite effect. These coffee breaks were support, not wasted time.

For those practices that are part of PLM, enabling rich, interpersonal communications should be encouraged, not discouraged. While it may be challenging to identify and separate support activities from wasted time, it is something that a successful implementation of PLM will require.

Change Capacity

The second issue with respect to people is the determinant of change capacity. PLM requires a substantial amount of change by people within the organization. A determining factor in the success or failure of PLM is this capacity for change. Even if people are excited about the prospect of PLM, the excitement will fade quickly if PLM taxes their capacity for change. The relevant factors regarding change capacity are as follows: the magnitude and timing of changes, the ability of constituencies to change, and the willingness of constituencies to change.[6]

The scarce resource that all people have in common in addition to time is attention.[7] People have only so much attention to give to their surroundings. If we want to convince ourselves of this, all we have to do is try a simple illustration. Drive a stretch of highway while talking on a cell phone. Then drive the same stretch not using the cell phone. After using the cell phone, we will have little or no recall of that stretch of highway. When not being on the cell phone, we will notice billboards, houses, and other surroundings. We only have a certain amount of attention available, and talking on the cell phone uses up a substantial amount, which we would otherwise use to attend to our surroundings.

Change also requires us to expend a substantial amount of attention. The more change that is occurring, the more attention that is expended. With respect to magnitude of change, if we are expecting people to make changes in the way they perform their task, we should not to be surprised that it is more difficult to make major changes than it is to make minor changes. We need to assess the magnitude of change that we are asking them to adopt. While this is a qualitative assessment, we generally have a good enough idea about our people to understand what they can handle.

However, when we look at change we cannot simply look at the magnitude of the change we are imposing on people. Change is a cumulative issue. Asking people to make a number of small changes can be more taxing than asking them to make a single change of greater magnitude. As managers, we tend to forget this when assessing the change capability of our people. We size up their ability to make change and then size the changes accordingly. However, we do not take into account that they are not only attending to the change that we have judged them capable of handling for our PLM project, but they are also attending to changes in the ERP system, the human resources system, and the customer resource system.

What appears to be a manageable change required by PLM is unmanageable by people when coupled with the other change initiatives our people are faced with. When we look at the magnitude of change that we are asking people to deal with, we need to look at the magnitude of change overall, and not just that of our initiative. It is therefore very important that we assess the timing of changes that people are faced with in order to phase these changes in such a manner that they do not exceed people's overall capacity.

However, it is possible to do something about people's overall capacity for change. The capacity for magnitude of change is like stretching a muscle: the more it is stretched, the more flexibility or capacity it develops. People in organizations that experience a great deal of change have more capacity for change than people in organizations that have a long history of doing the same tasks the same way. For the latter organizations, phasing in the changes to decrease their magnitude may be the only method for reducing widespread resistance to change. Introducing small changes in order to "stretch" the capacity for change should precede larger changes.

The next factor to consider is the ability of the constituencies to change. This goes to an earlier point that all people may have the capacity to change, but they may not have the capability. And so we need to assess whether they have the correct education, training, and support in order to be able to make the changes required of them. If we are asking them to perform changes but are not giving them education, training, and support, then we should not be surprised when those changes fail.

An example of this is a company that has always used a paper-drawing vaulting system and now implements a PLM vaulting application for math-based CAD designs. This change means that there is now no paper, and all designs are handled electronically. To adapt successfully to this new system, we need to make sure that the people are well trained and/or educated to handle the computer skills necessary to file, search, and retrieve their drawings. If we are going to successfully wean these people from paper, we need to make sure that we have special training and support as they make this transition, in order to assist them.

The willingness of constituencies to change is not to be underestimated. Many initiatives fail, not because they are flawed in their technologies or processes, but because the people who are using them are not interested in having them succeed. If the constituencies are structurally set so that they are not interested in making the changes, they feel threatened by the new system, or they simply are satisfied with the status quo, then their ability to make changes is severely limited.

The willingness of people to change is based on three major factors: their belief systems, their reward and punishment systems, and their available options.

The best way to get people to perform is to make what we are asking them to do consistent with their belief system. There are two aspects of their belief system: one is personal and the other is organizational. If a person has had successful experiences with change, that is if their personal belief system is that things get better with change, then they will be quick to adopt new ways of doing things.

However, even if that is the case, if the organization, or more precisely their peer group within the organization, does not believe in change then that person will be viewed as abnormal and pressured to conform to the rest of the organization. So for change to be successful, in addition to assessing an individual's ability and beliefs in what we are trying to accomplish with PLM, we need to assess the organization to which he or she belongs.

The second aspect is a reward and punishment system. It is well established that people's work habits are determined by their compensation system. So one of the best ways to get people to do what we want them to do is to compensate them for the behavior we want. While punishment systems are often effective at preventing behavior, they generally are limited in their ability to get people to perform new and positive initiatives.

What we would like to do is to craft reward systems that would encourage behaviors that are consistent with the new activities in PLM. An example of this might be a bonus system that would encourage design engineers to categorize drawings when they vault them so that every time a design is reused that engineer would get some sort of reward. Or we could craft a bonus system that scored manufacturing processes for process reuse. Above a certain score, the manufacturing engineer would be eligible for a reward. If we craft a reward system along these lines we can motivate people in the appropriate manner.

Finally, the last factors affecting change are the options people have. If we allow it, one of the strongest options is always no change or the status quo. One way to short-circuit this is to take the status quo off the table. By that we mean that people are not given the choice between using their old way of doing things and the new way. If they are given that choice, there will be a strong tendency for people to feel very comfortable with the status quo and not want to make the change.

So, in moving people to a new way of doing things, be it the engineering group or the manufacturing group, we must seriously consider making that change with no prospect of going back. While we need to assess this very carefully, leaving the status quo on the table can be very detrimental to the change process.

The next aspects of people that we need to consider in our approach to PLM are organizational ones. The two elements of organizational considerations are structure and the authority-enablement continuum.

With respect to structure, organizations need to be structured appropriately in order to encourage the flow of information across functional areas. If the organization is structured such that functional areas such as engineering and manufacturing are strictly separated and the flow of information is inhibited or only crosses functional areas at well-defined junctures, then the ability of an organization to share information across functional areas will be substantially limited.

In addition, if there is little experience with cross-functional teams, then organizations will tend only to focus on their own functional areas to the detriment of other functional areas. If an electrical engineering design group is functionally isolated with little or no communication with their counterparts in mechanical and programmatic design, then the electrical group will tend to do whatever it is that it thinks will lead to an optimal design for its discipline, irrespective if it is suboptimal at the mechanical and programmatic level.

On the other hand, organizations that encourage cross-functional teams of various kinds will see the benefits of cooperation across these various functional areas, and they will tend to get designs that optimize the design itself and not the particular special area that is being worked on.

The next element of organizational structure is authority. If the authority of an organization exists fairly high in the organization, then decisions about sharing information will only be made at that level. People will be concerned about overstepping their bounds of authority and will tend, in the absence of direct instruction to do so, not to share information. They will tend to develop their own domains of knowledge that will be closely controlled within their organizations, and, unless directed to do so by higher authorities, will tend to keep that information to themselves. In addition, what information they do share will be limited because of a fear that they will be sharing more information than the authority will allow them to share.

However, if the organization is one of enablement, then the participants in that organization will make decisions on the basis of the project at hand and will cross functional boundaries in order to obtain the results that they desire. If an organization feels enabled, then it will have a much greater tendency to engage in cooperation because it is not seeking authority for sharing of information or engaging in tasks that are not predefined by the authority structure.

A quick test of where the organization is on the authority-enablement spectrum is to ask whether it is better to beg for permission than beg for forgiveness. If an organization is high on enablement, people will be inclined to take the initiative. If it turns out to be the wrong thing to do, then they will need to "beg for forgiveness," which implies that forgiveness is not an uncommon thing. Conversely, people in a highly authoritarian organization need to "beg for permission." The consequences of getting caught showing initiative are much too severe to consider ever going ahead without permission. In authoritarian organizations, there is rarely any forgiveness.

Process/Practice

While "process" has always gotten the starring role in the "people, process, technology" triangle, practice deserves equal attention. There are a number of reasons why practice has not gotten much notice. The first is that, historically, information technology could only support processes. Information had to be highly structured and defined to fit the then-current capabilities of information technology. These capabilities were suitable only to support processes that were also highly structured and well defined, not practices that required unstructured, free-form information.

The second reason is motivational. As we discussed in the first chapter, we would like to believe we are only dealing with processes. Processes are well defined and better behaved than practices. We can flowchart a process. We can make an algorithm out of it. The inner workings of the black box are clear. Practices are much messier. The inner workings of the black box are opaque. We know if we provide the inputs, we generally, although not always, get the outputs. But we cannot always explain why. As one CIO of a Fortune 50 company said about practices, "We provide the information and then leave room to let the magic happen."

It is important to distinguish between processes and practices because this is where a source of major frustration and disappointment occurs. If we deal with a practice as if it is a process and attempt to make it more efficient by limiting "unnecessary" information and "extraneous" communication, we will generally get less efficiency, not more. We will also get managers frustrated that their carefully scripted systems do not give them the intended results and a disgruntled workforce that instinctively knows that its practices have been made worse and not better.

It is also important to understand that processes and practices can both exist within the same procedure or task. The process for a design change on a complex product may consist of specific steps that must occur each and every time and specific sign offs that must occur at certain specified points. However, deciding whether the change is consistent with professional and organizational standards in order to be approved or sent back for revisions is most likely is a practice. While calling the whole design change procedure a "process" is common, failing to understand that practices are involved and must be treated differently is also all too common. Attempts to "optimize" these practices almost always lead to failure and frustration. Processes and practices require different approaches.

Focus on Process

In spite of the confusion between process and practice, process is still a very legitimate issue for PLM to focus on. The more we can define processes, the more we can work toward increasing efficiency in a systematic fashion. Because it is much more difficult to define and measure practices, we should be analyzing our procedures and tasks and attempting to define and separate out processes from practices wherever possible.

With respect to PLM, we need to consider the following. First, we need to have a deep, not stylized, understanding of our processes. Second, we need to insure that they are explicitly defined and not tacit. Third, we need to re-engineer those processes for a digital environment. Finally, we need to integrate processes across the organization.

The most common example of a stylized process is one we learn in high school civics class. It is called "How a Bill Becomes a Law." It shows a nice flowchart of the purported steps that a bill moves through in order to become a law of the United States. The problem is that how a bill actually becomes a law bears only a cursory resemblance to this flowchart. While there are process elements present (e.g., moving from committee to floor, voting, and signing), anyone responsible for automating this "process" from the flowchart would find the application useless to the participants.

Unfortunately, a good number of processes are stylized in this fashion. Attempts to automate them meet with the same fate as our "How a Bill Becomes a Law" application. It is useless for the participants. To automate processes within the organization, business analysts must "get under the covers" and find out how the processes really work. Unless there is a deep understanding of how the processes really work, attempts to analyze and automate those processes will not succeed.

Closely related to the need to understand deep processes is the need for these processes to be explicitly defined processes. What is of express concern is that there are tacit aspects to the process that are supposedly deeply known. In many organizations there are tacit processes that come about as people try to find more efficient ways to do their jobs.

If the process is supposed to go from A to B to C, it may be that the person doing process A unofficially consults with people doing processes E and F because he or she knows that, unless he or she has that information, the work is going to come back for further revision. The process of one company is that the design engineers define the product characteristics, have them signed off by engineering management, and then send these specifications to purchasing for sourcing. What they really do is check with purchasing for the vendors that purchasing favors, and specify those products into the design.

What the engineers want to avoid doing is specifying a product that then goes to purchasing only to have purchasing send the design back for revision because the vendors that can provide this particular specification are either not qualified, not approved vendors, or have unacceptably long lead times. So even though the process would show the engineers fully defining the product, getting it approved, and then shipping it to purchasing for acquisition, the reality is that the process includes consultation between engineering and purchasing prior to the final specifications being formed and sent on to purchasing.

This is an example of an explicitly defined process that defines a tacit or unofficial process that exists in the organization. The issue is that these tacit processes, when an attempt is made to automate them, all of a sudden stop working because the people who are supposed to be in the loop on the tacit processes no longer have access to the information. As a result, what was once a good working system is now made to work the way it is formally defined. The result is that this system adds more work into the environment. What was once an informal networked process has become a formal, sequential, iterative process. The result is an unworkable or an inefficient process that mirrors the formal process but does not work.

To get the most out of PLM, processes need to be re-engineered for a digital environment. This is a very common issue because processes are built not only around the information, but around how the information is delivered. A lot of processes rely on the fact that they were built around paper flowing from person to person, and that the presence or absence of a paper would generate an action. In a digital environment, we want to eliminate the paper, not use it as a crutch with which we are continuing an old process. So even simple things such as signatures need to be assessed and enabled in a digital manner as opposed to relying on pieces of paper that are only an indication that the approval has been made.

However, in older organizations or in organizations with workers that are not comfortable in a digital environment, the lack of paper can cause great anxiety among the people doing the work. There may be an interest in continuing on with the paper system simply as a reassurance or backup. This will add additional costs and inefficiencies to the system and needs to be looked at carefully to see whether the paper or other material manifestation of information is necessary or not.

Finally, we need to integrate the processes across the entire organization. When product information was limited to a specific function, processes could remain independent and unintegrated across functions. However, if we are building the product substructure pipe and not confining the product information within silos, we need to look at the product processes across functional areas.

A change process that defines the change to a product but does not alert manufacturing to the change so that it can review the production processes, or does not alert service so that it can update the service processes will be inefficient and waste substantial amounts of time, material, and energy. Manufacturing will eventually become aware that there is change, and it will have to scrap material. Service orders the wrong replacement parts and will have to reorder the right parts after it finds the old parts do not replace the inoperable parts. What is a complete waste of time, energy, and material is designing a process for recycling and disposal, only to have that process be unavailable to the recycling facility when it is needed.

While there are other aspects to processes and their uses in eliminating wasted time, energy, and material, the ability to define deep, explicit processes, re-engineered for the digital environment and integrated across the organization, is the most pressing and challenging for PLM. Implicit in all this is the understanding that we have really identified what are processes and not lumped practices in with processes. Practices have their own issues, which we will deal with next.

Focus on Practice

As discussed in the first chapter, there is more to the people and processes relationship. The vast majority of information that we deal with on a daily basis is unstructured There is fuzziness or incompleteness in that information. Yet people are required to make decisions and act on those decisions. They do so through their practices. Unlike processes, practices are non-algorithmic judgmental activities. Practices use both deductive and inductive reasoning. Practices require the participants to discern the patterns in a pool of seemingly unrelated data and information.

The information requirements are different for practices than those of processes. With processes, the focus is on the movement from state to state as quickly as possible. With practices, the focus is on collecting data and information at each state in order to build a pool of information to improve the recognition of patterns in the future. However, when processes and practices are both part of a task or procedure that is being automated, the information goals of practice often get only minor attention.

An example of this is engineering change workflow. The focus of applications that support this task is the movement from state to state (e.g., requirements identification to requirements approval to specification development to design change to testing approval to engineering management approval). The information about how decisions were made with respect to design alternatives, approval of one design over another, or the selection of testing methodologies is usually treated as a free-form entry and is gone when the transaction is completed. These entries could add to the pool of information that could improve future decision making.

For PLM, the goal is to provide the pool of data and information and assist in discerning the correct patterns. For PLM, the important factors regarding practice are providing standards and guidelines, capturing and categorizing exemplars, and providing rich interpersonal communications and coordination. Finally, the practices of PLM are, to a large degree, defined and/or supported by educational and professional institutions and cannot simply be controlled by the organization.

The ability for computers to handle unstructured data has improved dramatically over the last decade. Search engines now continually catalog the contents of the World Wide Web. Semantic searches where the context of what we are searching for is discerned by the search engine may not be far off.

It is now feasible to create standards and guidelines that can support practices. Standards and guidelines that once existed in paper form and were useable only by experienced experts can now be made available on demand to those less experienced. In addition, the standards and guidelines can be linked to design and testing information to provide an understanding of the rationale for design and approval decisions.

Capturing and categorizing exemplars of practice is also a function that PLM can enable. This will increase the pool of information that will enhance judgmental decision making. Decisions that can be reduced to unambiguous rules belong in processes. The decisions that cannot be reduced to unambiguous rules belong to practices. As discussed earlier, the decisions involved in practices have as their goal relatively clear objectives or outputs. However, the inputs required and the mechanisms by which those inputs are considered to produce the decisions are fuzzy, and often not well understood.

The reason that this may be the case is that human beings are superb at comparison and pattern recognition. However, in order to be able to do this, they need information to compare against and patterns that can be recognized. The more judgmental the decisions, the more information required. PLM can assist in practice by categorizing and providing exemplars of past decisions. If we have these exemplars available, we can use them to compare against our current decision.

If designers have exemplars of designs that were both approved and rejected, they can use these to make decisions about their designs. If manufacturing engineers have exemplars of successful and unsuccessful manufacturing processes, they may be able to find commonalities with their proposed processes and weed out ones whose patterns are similar to failed processes.

PLM can assist in the capture and categorization of exemplars, but expectations about its impact need to be realistic. It is commonly stated that engineers spend a substantial portion of their time searching for information. The implication is that this is wasted time that could be saved. However, a substantial part of that search time is looking for and at exemplars. Engineers are looking for designs that contain patterns that are comparable and compatible with their current requirements. In point of fact, they really do not know what they are looking for, but they want to see as many exemplars as possible.

PLM can assist in this—however, not in the way expected. PLM will not reduce this search time by presenting the single instance of information that the engineer is seeking, but by providing him or her with exemplars that are relevant to the search. Wasted engineering time can be reduced, but not in the way commonly thought of when people see this statistic on wasted time. In fact, it may be that engineers should spend more time in "searching" (i.e., looking at exemplars), rather than less time.

In addition, practices require rich interpersonal communications and coordination. Unfortunately, this is sometimes the opposite of what information technology systems attempt to do, in that some information technology systems attempt to eliminate or reduce the amount of information that is communicated, in the name of efficiency. However, practices require these richer interpersonal communications and coordination.

As mentioned before, "When the going get tough, the tough go to coffee," is basically an indication that people are looking for assistance in order to be able to make sense of the myriad inputs in the context that they're seeing them in to then be able to get to the final outputs. As a result, they need more information, not less information. They need to make the decisions about what information is relevant and not relevant, and they need to reorder and reprioritize that information as they work their way through the problem.

A noted CIO describes one of the PLM applications as a "chat room for engineers." He is not far off in his assessment. PLM, with its focus on capturing and categorizing information, should act a "chat room." The technicians that went for coffee were clearly in a chat room. The problem is that verbal chat is ephemeral. It exists for the people in hearing range, but quickly dissipates. PLM needs to facilitate the rich communications of chat rooms, but capture and categorize those communications so that they do not require the user of that rich communication to have been in a specific place at a specific time to benefit from it.

Finally, practices are defined and supported by educational and professional institutions and associations. The professionals engaged in practice, such as engineers and sales and customer support personnel, learn their basic practices at educational institutions. Professional practices define how they view the work that they go about doing. As a result, it is often difficult to change practices within an organization if they are not consistent with the practices learned by these professionals in educational professional institutions.

For example, if we radically change the definition of good engineering practices, we find that our engineers will unconsciously balk about performing those practices or attempt to modify them so that they correspond to the practices that they have been taught in schools. With respect to Product Lifecycle Management, this means that the cross-functional approach that is required by Product Lifecycle Management must be moved down into the educational and professional institutions.

As the educational institutions are structured today, students are not only being focused on specialty areas, but on subspecialty areas. So, not only is an engineering student focused on electronics, she may also be focused on microelectronics. As a result, this idea of sharing or having an understanding of the cross-functional requirements may be difficult to achieve because engineers are taught that the focal point of their professionalism is in their subspecialty. Attempting to optimize designs around their subspecialty is what they're trained to do. However, this may suboptimize the product in general, and we will have to educate these students to have a wider look at the world rather than a narrower one.

However, practices are generally pervasive throughout the organization because engineers, for example, no matter which area they specialize in, perform the practices that exist to determine whether their designs are good designs or bad designs. And generally these practices will be the same, irrespective of the organization they go into, although there may be individual idiosyncrasies in specific areas. These are practices that engineers bring them with them as they move from job to job, as opposed to individual processes that may change from organization to organization.

Technology

Because the capability of the information technology aspect of Product Lifecycle Management is heavily dependent on the applications that solution providers develop in this area, any discussion of these applications would quickly appear dated. The companies that have been responsible for the development of PLM applications have continually undergone change. Acquisitions, mergers, and spin-offs have been commonplace. Even the names of the applications have changed, so that references to specific PLM applications would be familiar only to those individuals who had long-standing experience with the industry.

What we can say about the PLM application space is that there are a substantial number of highly capable software developers who view PLM as their primary focus. They continue to expend substantial research and development resources on new and innovative software and to push the boundaries of recording, managing, retrieving, and using product information. Since PLM is in its infancy, we will see the choices of applications increasing and becoming more robust.

Since there is a wide variety of applications targeted at different aspects of PLM and different industry uses, organizations that are in the process of selecting PLM applications need to assess the functions that they require against the functionality of the products in the marketplace, both current and planned. To help with these decisions, there are a wide variety of research firms and consultants who are well versed in PLM and can help with an analysis and assessment of PLM applications.

There are some observations and issues regarding technology that are independent of any particular application and would be relevant to any PLM implementation. These issues pertain to the environment that PLM operates in, or considerations that any PLM application must take into account. These issues are

- PLM needs an adequate technology infrastructure.
- PLM applications should be open and harmonize with other applications.
- PLM applications must be configurable and not customizable.

- PLM applications must be useable and embedded.
- PLM applications must be utility-like in their performance.

The Infrastructure Hurdle

Infrastructure is one of those issues that is often overlooked because we make the assumption in today's digital world that adequate infrastructure exists. However, this is not always the case. PLM, with its requirements for access to substantial amounts of math-based designs, can put a significant strain on infrastructure that is not sized to handle it.

Infrastructure is a hurdle issue, meaning that if we have excess infrastructure over and above what we require, we do not get any benefit from this excess. However, if we have a shortage of infrastructure below what we need, that can cause slow adoption and even failure of the project. If we do not have adequate computing capability and communication bandwidth and/or storage capacity such that the users of the system find that their ability to do their work is degraded, then the PLM system will not succeed because people will either work around the system or circumvent it entirely. At a minimum, their complaints will be at a level such that their managers will have second thoughts about their support for PLM.

As a result, a careful assessment of infrastructure, given the new and substantial demands that Product Lifecycle Management makes on it, should be one of the first things undertaken in assessing the ability for an organization to adopt Product Lifecycle Management. The major equipment vendors such as IBM and Hewlett-Packard are attempting to address these issues with things like computing on demand for IBM and adaptive computing for Hewlett-Packard. Their intent is to ensure that the infrastructure has the capability, or the capability can be added on demand, so that the infrastructure is scalable with the requirements of Product Lifecycle Management.

Open, Harmonized Applications

There is such a variety of product information and its uses that it is difficult to imagine that a single solution provider can encompass the entire breadth and width of PLM. As a result, we should expect that different solution providers will focus on and develop superior products in selected areas of expertise. Since it unlikely that standards will be developed that will encompass all the requirements of PLM, we should look for openness and harmonization in their product offerings.

By openness, we refer to the ability to understand and use the information from their applications at some level of granularity. For example, we would like the capability of visualization and simple manipulation of math-based CAD designs in different CAD system formats, even if we needed to use the native application to make changes.

By harmonization, we mean we would like to have different applications not be incompatible with one another so that we can transfer information from one application to another for different uses. For example, we would like math-based CAD designs from any CAD solution provider to be useable by Digital Manufacturing applications in factory floor simulations. While the applications might not have their formats directly used by one another, harmonization might be enabled through the use of common exchange formats such as XML.

Configurable, Not Customizable

Customized software is the bane of any information technology support organization and is especially problematic for new initiatives like PLM, where applications with new functionality are developing rapidly. Any customization of such applications is a prohibitively expensive undertaking, as it will delay adoption of new and improved applications because any customization will have to be ported and tested to the new application at a substantial cost in time and effort.

Also, organizations should not be taken in by claims that the new applications represent "best practices" and that the organizations should change their processes and practices of operating in order to implement the application in an uncustomized state. As described in Chapter 1, "best practices" is at best a dubious claim. The application may implement "healthy practices," but it needs to accommodate the "best practices" as each organization has defined and developed them for its specific situation.

Configurable applications are the solution for applications that are not customized but are tailored to the organization that acquires them. With configurable applications, we can implement our "best practices" without writing new and expensive code. As an example, General Motors used to have 100,000 lines of custom code in its Product Lifecycle Management systems. It has reduced this to under 20,000 lines of code through the incorporation of configurations in the software that it acquires. This does not mean that General Motors has changed the way that it does business, nor does it mean that GM has dropped the processes that it feels gives it a competitive advantage. Instead, what it means is that the software is now configurable so that GM can get the desired results without the problems and expense that software customization entails.

Embedded and Useable

It is imperative that the applications reflect the way that people do their jobs. Too often, technical consideration of software development overrides the usage requirements of the application users. Software developers architect the application because of how the underlying database is organized or the technical requirements of updating multiple databases. The problem is that this is not how people use the information. The result is that people must update the application in a manner that does not reflect their job processes and practices. This wastes time, energy, and material.

The core problem is that we align our information technology with our business processes and practices where we should be embedding them. When we talk about alignment, we talk about roughly coupling our information technology with the information requirements of the people using that information. Some misalignment is to be expected.

However, we need to be talking about embedding. With embedding, the information system is an integral part of the job processes and practices. The processes and practices and the software application need to work as one seamless unit. The application must reflect how the job is done, not how the programming is easily accomplished. Applications for PLM must be eminently useable and embedded into the job. Alignment will only mean that we have put the software developers' needs over the needs of the application users.

Utility-Like Performance

If we want our people to capture, retrieve, and use product information such that it is embedded in all their processes and practices, then we need to make that capture, retrieval, and use ubiquitous to them. Much as they do not give a second thought to plugging a power cord into the wall to get electricity or switching on a lamp to get light, people need not think about obtaining and using information.

This means that the applications that support PLM must work each and every time the user "switches" them on. If our people have to go through periods of debugging each time a new release of the application is installed, they will have second and third thoughts about trusting their valuable information to such an application.

At a minimum, they will have alternate means of handling this information, such as paper forms or personal spreadsheets that will waste time, energy, and material. Where information is peripheral to their jobs, people may tolerate imperfect software. However, for PLM to be fully adopted and embedded into the organization, its applications must have the reliability that we expect from any other utility that we use.

Summary

In Chapter 5, we have discussed the people, processes/practices, and technology that comprise the approach of PLM. With respect to people, we discussed the impact experience, education, training, and support has on people's ability to adopt PLM. We took a close look at an important element in moving to PLM—people's capacity to change. We discussed the magnitude and timing of change, the ability of people to change, and, very importantly, their willingness to change.

We discussed processes and the often-overlooked practices issues as they pertain to PLM. With processes, we focused on having deep knowledge of the actual processes, the need to have them explicitly defined, re-engineered for a digital environment, and integrated across the entire organization. For practices, we discussed providing standards, guidelines, and exemplars, along with providing rich interpersonal communications and coordination. We looked at practices as being influenced by education and profession.

With respect to technology, we focused our attention on the infrastructure hurdle and open, harmonized, and configurable applications. The technology needs to be useable and tailored to the people, not vice versa. It needs to be embedded into the organization and not aligned, with performance that is utility-like.

Notes

1. In this chapter, I am following the conventional explanation that an approach or system is "made up" of people, processes, and technology (with my addition of practices). The reality is much more complicated than that. These elements come together and transform one another. People (or more precisely, the ways people think) are not the same after their interaction with technology. Technology gets transformed as people use it, sometime very differently from what the developers intended. Processes and practices transform and are transformed by the people and technology. For those not satisfied with the more simplistic view of the interaction between people and technology, see B. Latour, *Pandora's Hope: Essays on the Reality of Science*, Cambridge, MA: Harvard University Press, 1999, especially Chapter 6.

2. While records of business transactions date back thousands of years, true information technology that developed methods and means of organizing, recording, and retrieving information in order to control organizations developed over the last century or so. See both J. Yates, *Control through Communications: The Rise of System in American Management*, Baltimore and London: The John Hopkins University Press, 1993 and J.R. Beniger, *The Control Revolution: Technological and Economic Origins of the Information Society*, Cambridge, MA: Harvard University Press, 1986.

3. See S.J. Spear, "Learning to Lead at Toyota," *Harvard Business Review*, 2004, *82*(5), 78.

4. While we are touching on some elements of education, training, and support with respect to PLM, this area is of intense interest to organizations in general. Specifically, managers have great interest in the idea of creating and maintaining an organization that continues to learn, where education and training are not point events. See P.M. Senge, *The Fifth Discipline the Art and Practice of the Learning Organization* (1st ed.), New York: Doubleday/ Currency, 1990.

5. See J.S. Brown and P. Duguid, *The Social Life of Information*, Boston: Harvard Business School Press, 2000.

6. I expanded upon the ideas in the excellent book by C. Benko and F.W. McFarlan, *Connecting the Dots: Aligning Projects with Objectives in Unpredictable Times*, Boston: Harvard Business School Press, 2003.

7. For a discussion on attention as a scarce resource for people and their organizations, see H.A. Simon, "Designing Organizations for an Information-Rich World," in M. Greenberger (Ed.), *Computers, Communications, and the Public Interest*, Baltimore: The John Hopkins Press, 1971, pp. 37-52.

Collaborative Product Development— Starting the Digital Lifecycle

ALTHOUGH PLM CONCERNS the entire product lifecycle, it only makes sense that there is a focus on the early stages of the lifecycle to begin developing and implementing PLM techniques and applications. This chapter will focus on how organizations are applying PLM in requirements analysis and planning, design, and engineering. Since a great deal of the entire product's cost parameters are defined during these stages, a focus on these applications can benefit organizations immediately and give organizations experience at understanding the problematic issues that arise at the intersection of functional areas.

What Is Collaborative Product Development?

Collaborative Product Development or cPD[1] begins the first phase of the product's lifecycle. However, products do not appear out of thin air. Even products in brand-new product classes are based on, or components even borrowed outright from, older product classes. The automobile was a "horseless carriage." The Wright brothers were bicycle mechanics, and their revolutionary flying machine employed some components based on that and other existent products. The Xerox copier had components based on extant mimeograph machines. In most product development, new products rely on some aspects and/or components of products that have already been developed, tested, and used. In this respect, even the most novel products are collaborative.

Product development is driven along one of two lines: *customer driven development* or *entrepreneurial driven development*. In customer driven development, a customer specifies to a greater or lesser degree the requirements, functionality, geometry, specifications, and characteristics of the product. This allows for a wide spectrum of possible activities for the product developer.

At one end of the spectrum is the contract manufacturer. It receives completely defined product geometry, specifications (possibly even including process specifications), and characteristics. The contract manufacturer is responsible for turning these bits into atoms. At the other end of the spectrum, the customer simply gives the product developer the requirements. The product developer is then responsible for turning these requirements into geometry, specifications, and characteristics. In some cases, this includes developing entire subsystems, where the product developer becomes the customer to its suppliers.

Entrepreneurial driven development is simply an extension of customer driven development. In this case, the entrepreneurial product developer also develops the requirements. The entrepreneurial product developer makes the sometimes risky assumption that the requirements that it designs for will match the requirements of some set of potential customers. In cases where new or novel technology is utilized, the product developer may be pleasantly surprised that the potential customer pool greatly expands as the technology increases its functionality.

In both the customer driven and entrepreneurial driven cases, the common theme is that there is collaboration between customer and product developer around a common understanding of product requirements. This collaboration demands that requirements be translated into a single representation. This single representation must be capable of being turned into one or more physical products. The chain of events is ideas to bits to atoms.

Based on this view of product development, we would define *Collaborative Product Development* (cPD) as follows:

Collaborative Product Development, an initial stage of PLM, is an approach to capturing, organizing, coordinating, and/or controlling all aspects of product development information, including functional requirements, geometry, specifications, characteristics, and manufacturing processes in order to provide a common, shared view as product requirements are translated into a tangible product and to create a repository of product information to be used throughout the product lifecycle.

As the beginning of the product lifecycle, Collaborative Product Development begins building the Info Core in the PLM Model and the image of the product-to-be in the Information Mirror Model. A common shared view, meaning singularity of information, cohesion across different views, and traceability of the product design, enables the collaborative aspect of product development.

Collaborative Product Development, as a part of PLM, captures and organizes the information about the product as requirements are being transformed into a realizable, physical product. The information that is captured and organized can not only be used during the product development phase, but, by being captured and organized, can then be used in other phases of the product's lifecycle, including manufacturing, usage, and disposal.

As a part of PLM, cPD coordinates and/or actually controls this information during the development phase so that product developers can trade this information for the wasted time, energy, and material that they might otherwise spend working on outdated versions, reconciling incompatible product views, or recreating product information that already exists. By creating a repository of information, the product information becomes an asset of the organization that can be used not only for the current product's lifecycle but in the development of new and different products.

Although authoring or CAD tools are an important aspect of this phase of PLM, Collaborative Product Development is composed of more than tools. As we have done consistently, we refer to Collaborative Product Development as an approach that requires people, process, and practices, as well as technology such as authoring tools, workflow management, and search engines. We will discuss the different aspects of Collaborative Product Development that organizations are focusing on in order to use PLM to drive their product development processes and practices.

Mapping Requirements to Specifications

One of the most useful functions that Collaborative Product Development can perform is developing and mapping the product requirements to its specifications. As noted above, the product requirements are what drive the geometry, specifications, and characteristics. Requirements are such things as

> Transition times between conventional and vertical thrust and back again for a jet fighter with short takeoff and landing (STOL) capabilities must be accomplished consistently in one to three seconds.
> The pacemaker must mimic the heart's natural rhythm by adjusting the rhythm according to a person's activity level.
> The stopping distance of the automobile going 100 kilometers / hour must not exceed 40 meters on dry pavement.

However, once these requirements are translated to specifications and the specifications modified during the design processes, the linkage between requirements and specifications can be lost. The focus moves to the specifications and, if the linkages are not maintained, the design engineers lose their ability to discern which specifications have degrees of freedom and are changeable and which specifications are unchangeable in order that certain requirements are met.

As the product development phase progresses, it is not uncommon to experience "function creep." Function creep occurs as the product design takes shape. It occurs irrespective of whether the development is customer or entrepreneur driven. In customer driven creep, the customer decides that it needs or wants additional functionality as it gains a better understanding of the product.

The only defense against customer function creep is good information. In the spirit of customer service, product developers will sometimes agree without thinking to what appear to be simple requests. However, sometimes these simple requests turn out to be anything but, since they impact some other functionality constraint. Without access to that information, product developers can find themselves taking on costly and time-consuming additional work. While it may not be possible to refuse customer driven requests, armed with the right information, product developers can at least understand the full ramifications and price the requests accordingly.

Entrepreneurial function creep is sometimes more insidious. There are generally two sources of this function creep. The first is that the product developers themselves get enamored with their own capabilities. They have functionality A, but with some clever engineering they can add functionality B. Since product developers view themselves as clever engineers, functionality B gets added whether customers might use it or not.

The second source of entrepreneurial function creep is the sales and marketing group. Even though it has signed off on the original functionality, it is back saying, "Things have changed. If we don't have this additional functionality, we don't have a product people will buy." This is a difficult situation for organizational leadership. It does not know whether this is the case or whether this is the sales and marketing group's way of developing an a priori excuse for not meeting its sales targets. The decision process of whether or not to add the new functionality is the subject for a different book. Understanding the costs and implications in order to be able to make an informed decision is very relevant to Collaborative Product Development and PLM.

Regardless of where the request to add additional functionality arises, the information that cPD captures and organizes is required in order to understand the full ramifications of adding additional functionality. Building and maintaining the links between requirements and specifications are crucial when adding new functionality with its ensuing specifications. Adding an inch of wheelbase to increase a car's roominess adds weight that affects the size of brakes necessary to stop the car in the required distance. Adding a heavier, more powerful missile to the weapons payload of a jet fighter changes the amount of engine thrust required to accelerate the plane to a required speed in a required time.

It is important that there be one source or singularity of information and a consistency across views. Any change to specifications mapped to functionality must be examined and reconciled. Collaborative Product Development, with its focus on mapping requirements to specifications, is required to prevent the need to reconcile functional trade-offs far into the development phase where it is much more costly to remedy them.

Part Numbering

Part numbering would appear to be a simple, innocuous aspect of product development unrelated to PLM, but it is not. Part numbering is a key issue in abstracting product information in order to simplify our handling of that information. By assigning a part number and mapping it to a specific and, hopefully, unique set of specifications about a product, we are attempting to simplify information processing and communication across the organization.

By using a part number, we eliminate having to transmit all the product specifications and characteristics. We also eliminate the need for the recipient to carefully examine those specifications and characteristics to understand which specific product among the group of possible products matches up.

This works well when there are a limited number of products, when products are simple, and when everyone working with the product has exactly the same specifications and characteristics. This situation is equivalent to singularity of information. However, as we have seen, the number of products continues to proliferate, their complexity is rapidly increasing, and, with growing scale and globalization, it is difficult if not impossible to have a technology-unassisted shared understanding of specifications and characteristics for different groups.

The result is often that a single part number has similar but not exact specifications and characteristics. This difference, even if seemingly minor, can cause a major waste of time, energy, and material because this difference requires reconciliation and remediation when it is discovered.

One solution to this problem of a lack of shared understanding of the specifications and characteristics is to limit the scope of those the shared understanding has to apply to. As a result, engineering has its own part number, manufacturing has its own part number, and service develops a new set of part numbers for its use. Each functional area develops the specifications and characteristics that their function needs to deal with. This solution may solve or at least diminish the problem within the individual functional areas. However, the results are clearly the siloing of product information and the difficulty of piping that information from one function to another.

Another solution to this problem of an increasing lack of shared understanding of the specifications and characteristics is to proliferate part numbers. When there is doubt as to whether everyone shares a common understanding, simply assign a new part number and move on. Part number proliferation brings its own set of problems and new sources of wasted time, energy, and material. There is the cost and confusion of maintaining information on products that are the same except for their part numbers.

But the significant waste of part number proliferation occurs as the product moves through its lifecycle. There can be, and usually are, substantial wasted costs in sourcing, acquiring material, stocking inventory, and stocking service parts for the duplicate product. Creating a separate part number creates an independent path of duplicate costs that are unnecessary and wasteful.

What PLM brings to bear on the problem is its ability to enforce singularity of information. If there is singularity of information that is accessible to everyone, then a part number maps to a specific set of specifications and characteristics. If two parties require different specifications, then they are forced to resolve their differences immediately, either by compromising on a common specification or by creating a new part number.

PLM will not inherently stop the proliferation of the identical specifications and characteristics with different part numbers. However, good processes that require, among other things, a search for parts meeting the specifications and characteristics should minimize duplication, along with fostering reuse as discussed in the next section. Also, we should expect that eventually the technology can compare new parts with its current repository of parts to ferret out duplicates.

"Smart part numbers" are an attempt to encode information about the product characteristics into the part number. Smart part numbers have built into their structure information that identifies different aspects of the part. So, for example, the first three digits of the part number might refer to the family of products that the part is from. The next three digits might refer to the model number, and the next two digits might refer to the revision.

Smart part numbers are intended to allow us to quickly categorize the component or the product to which it is attached. They attempt to allow us to recall information that we can store in our memories so we can make simple assessments and decisions, such as what color it is, what engineering area is responsible for it, what vendor supplies it, etc. This is intended to save us the time it would take to physically acquire this information from an external source, such as a book, file cabinet, etc.

Smart part numbers are a second-best solution to the requirement of accessing information about the product or part by possessing only the part number. The reality is that smart part numbers are generally obsolete fairly quickly because the part numbers are rarely big enough to accommodate all of the various permutations that develop in the product. In addition, different functions use different aspects of a smart part number, depending on their interest in the categorization. As a result, smart part numbers may change across functional areas, leading to confusion as individuals make determinations about the part number based on their functional area, such as engineering or manufacturing.

What Product Lifecycle Management's info core allows to happen here is that the part number has no meaning itself but simply acts as a pointer to the always accessible product information. The implication of smart part numbers is that the underlying information about the part number or product number would be difficult to come by, or requires a fair amount of wasted time in searching and retrieving that information. If the information is readily available any time and anywhere and accessible with the part number, then accessing the info core will give anyone all of the information that he or she needs, without having to rely on the codes of smart part numbers.

With the implementation of PLM, smart part numbers are generally replaced by sequential part numbers. The users of the part number rely on their capability to easily retrieve the mapped specifications or characteristics and obtain far more information than would be available with smart part numbers. In this case, the trade-off is not atoms for bits. But having a small number of bits (part number) easily translates into having a large number of bits (specifications and characteristics).

Engineering Vaulting

Engineering vaulting is often an initial PLM project because it is so obvious and so necessary to organizations. Simply identifying and consolidating all the engineering specifications and characteristics in one repository is a tremendous benefit to organizations that lack such a central repository. As discussed earlier in the book, the rise of CAD systems allowed for the duplication and dissemination of product information, regardless of whether or not there was a process to control it. While this increased the usability and accessibility of this product information, the discipline of centralizing the information in drawing vaults was substantially eroded.

Since this CAD math-based data was available somewhere within the organization, and in many cases in multiple places within the organization, the perceived need to centralize the drawings degraded. However, this proliferation of duplicate and often inconsistent product information created substantial problems and inefficiencies of its own. People began to work on different versions of what they thought was the same product information, only to have to reconcile or redo their work when they discovered the inconsistency.

However, it is not enough simply to vault the product information. At the time of its vaulting, the product information needs to be categorized so that it can be retrieved. Some categorization, particularly geometric features, can be built into the design process. As

CAD application software evolves, some of these geometric features can be determined and self-categorized by the software. Features such as inner diameters and outer diameters are examples of such features. The software may even recognize and categorize simple parts, such as bolts, screws, and fasteners.

Other, more complex parts require human categorization. Such categorization is often dependent on the practice that has been established within the organization rather than on any absolute or objective criteria. The same part may be categorized differently in different organizations depending on their usage of the part and their previous practices. In order to create an effective vault, the key to categorization is consistency. If organizations can create vaults where they can consistently categorize the part or product and then consistently retrieve it on demand, these organizations can create a valuable asset that they can use to trade off for wasted time, energy, and material.

Product Reuse

Once engineering vaulting takes place, then other uses of this information can be made. The use of interest in cPD to most organizations is the ability to reuse this product. In a number of organizations, the easy answer to a functional requirement for a part is simply to redesign it again. With the information difficult to find and categorization nonexistent, engineers find that the simple expediency of redesigning the part solves the problem of having to conduct an ad hoc expensive search for a product that can be reused.

What organizations want to do is to maximize the use and minimize the duplication of parts that have already been designed. Part duplication and proliferation throughout the organization is probably the single most expensive source of inefficiency within an organization. Not only do these new parts need to be designed, but, like the same parts with duplicate part numbers, they then need to be sourced, inventoried, tracked, and serviced over their lifetimes. Collapsing multiple duplicate parts down to a single part is a tremendous source of cost savings for most organizations.

The idea behind part reuse is that engineers, when faced with a functionality requirement, will first search through the parts repository system to determine if a part that will fulfill this function already exists. For simple parts like bolts and fasteners, this will be a relatively straightforward process. Although, even here, there will be some need not only to look at the functional requirement for the specific purpose that the product is being designed, but also to look in the geographical neighborhood for similar parts that can be made consistent.

Thus, for example, if there is a requirement for a bolt of a specific length and a specific size, given the attachment requirements, the engineer might come up with one specification. However, if there are other bolts in the area that have slightly different specifications, the question ought to be asked, "Can I change my functional requirement such that I can use the same bolt in the same neighborhood so that repair and replacement would be much simpler?"

Unfortunately, today we find situations where there are a number of different fasteners in the same location that are all different. There is no reason why a common fastener could not be used, which would simplify the repair and replacement situation for when the product is in use.

While the reuse of simple and common parts is straightforward, the reuse of more complex parts, components, and subsystems needs additional attention. These reuse items are oftentimes part of an interrelated system and so are designed "in context." Simply picking up these designs and inserting them in a new product is not enough. Especially where these designs have inputs or outputs that are safety related, the engineers intending to reuse these designs must have a complete understanding of all the interactions and possible states this design can take. This is to ensure that the design has all the inputs it needs and that all the outputs it produces are accounted for and handled. Where informal practices that have been developed have been confined to a local group, the globalization of the design effort will require that these practices be captured and maintained with the product information. Having the geometry of the design will not nearly be sufficient.

Start and Smart Parts

Closely related to parts reuse is start and smart parts. These are a PLM mechanism to create an asset of the organization by capturing the forms and rules used in components of the organization's product. *Start parts* are prototypical forms of the building blocks that are used to create new products. Fasteners, gears, and housings are some examples of start parts.

Start parts are complementary to reusable parts. If the part can be used in its entirety, then parts reuse is highly efficient. However, if the part cannot be reused in its entirety, but must be modified, the modifications necessary to take the part back to a form that can be built on for the new part can be highly time consuming.

Start parts are designed for that purpose. As prototypical forms, they have the basic geometric shapes and characteristics that any part in this class must have. Based on this start part, designers and engineers can then add the specific form and characteristic changes necessary to meet the functional requirements that are needed to produce the new design.

In addition, as discussed in the last chapter regarding practices, start parts serve as exemplars for designers and engineers who may be new to the organization or who may not be familiar with the specific class of part. Rather than try to absorb a detailed standards manual concerning the part class in question, they can examine the prototypical part itself. This shows the elements of the part visually and in context. It is much more efficient than reading a standards manual and trying to understand all the relevant relationships.

Smart parts are also complementary to parts reuse. Smart parts are parts that are not only fully approved and working parts, but they have modification rules built into them. When the part needs to be scaled for a different use, the intelligence of how it scales is built in. As a simple example, if a bolt that is designed to support a certain weight needs to be longer, it will also increase its cross-section in order to maintain the same weight rating.

General Electric has used smart parts on far more complicated parts, such as those that make up jet engines. Very sophisticated rules about heat exchange, tensile strength, and minimum and maximum nozzle diameters are embedded in their smart parts. If a modification attempt is made, the smart part checks its rules to ensure that the new configuration conforms to the design requirements.

Start and smart parts are an excellent example of Lean Thinking. The information captured and embedded in start and smart parts is traded off for wasted time, energy, and material. The information from the organization's best designers and engineers can be captured and reused over and over again.

Engineering Change Management

Engineering change management is one of the most vexing issues within most organizations. While "best" practices would have us freeze the design early in the development cycle, the reality is that most organizations don't have the luxury of doing this. If the change requirements come from the customers, then they have little or no choice but to make the changes. However, even if that is not the case, there generally is additional information gained while the product is being designed that requires changes to be made. Assumptions about the resulting functionality of a specific design feature are proven to be flawed or suboptimal. Therefore, most organizations have to have a change management system in place in order to implement the changes within their design organization correctly and efficiently.

However, this is not simply an issue for engineering. As engineering changes occur, there are implications for both purchasing and manufacturing and service as the functionality and specifications of the product change. As these changes occur later into the release cycle of the product, the importance to these other functions of understanding the changes that are occurring increases dramatically.

The issue with a number of organizations, even when they have excellent change management within engineering, is that the information about these changes is not conveyed to the other functional areas in a comprehensive or timely fashion. In a number of organizations, this information is batched up and conveyed only at specific intervals, meaning that valuable time is lost in other functional areas, such as purchasing, manufacturing, or service, to react to or even have visibility of the changes that are occurring.

The Product Lifecycle Management approach that ties all of these functional areas into a common repository of information is required in order to process engineering change requests efficiently and effectively. Especially for those organizations where changes occur later into the development cycle, the requirement that they have visibility as quickly as these change requirements are known is of paramount importance. This is so as not to waste time, energy, or material on working on older versions that will need to be redone.

Collaboration Room

As the trend to globalization continues to accelerate, the need for effective means of collaboration continues to become more and more important. As organizations embrace Design Anywhere, Build Anywhere, and Service Anywhere (DABASA), they will need to create the equivalent of the project collaboration room in virtual space. Under a philosophy of cPD, we have already created the math-based product design in virtual space. This information is theoretically globally available via the Internet. All we need to do is create a window from the real world into the virtual world and the tools to manipulate the math-based product information.

Collaboration rooms are intended to do just that. They are also a way of insuring singularity of information—a key characteristic of PLM. While digital copies may be "checked out" for independent work, the shared understanding is that the designs in the collaboration room are the controlling version. It is an electronic version of Gustav Eiffel's wall. The requirements for the collaboration room are that it visually presents the product simultaneously to all participants (at least under the Grieves Visual Test condition); that it contains a project board that lists the development steps, status, and step responsibility; and that it captures the changes to the product as the development progresses.

Since the proposition to collect the development team in one physical location under the global development strategy currently employed by all large manufacturers is a costly and time-consuming one, the collaboration room must provide that capability. While asynchronous access—the equivalent of various participants walking in and out of the empty project room to work on the product—has a role in product development, pulling together the entire development team to work on a shared idea of the product and a shared understanding of the tasks and issues involved in creating it is critical to the efficient practice of product development. In addition, this collaboration has a better chance of fostering innovation and improving quality—key internal drivers of PLM.

The ability to have access to the requirements, specifications, and characteristics of the product is not enough. The requirement to communicate simultaneously in visual geometry is indispensable if we want efficiency. At a minimum this simultaneous, visual communication must meet the Grieves Visual Test, where we can inspect the evolving product from any angle and assemble and disassemble the various components.

However, for collaboration rooms to be truly effective, they will need to evolve to meet the Grieves Performance Test so that the participants can not only see the product, but put it to any and all functionality tests. The communication between participants can be voice or instant messaging text (which is gaining great popularity for this type of task).

The project board already exists in the form of the workflow management tools that exist within all cPD applications. Task assignments, task statuses, routings, sign offs, and other product development information are generally one of the first implementations of PLM. It is much better than a passive project board in that these PLM applications are proactive. They notify participants of task assignments and deadlines and alert product development team members of new requirements, changes, and missed deadlines. As a further benefit, all product development activities can be aggregated to give senior leadership a comprehensive view of all the development activities within an organization.

The last requirement of the collaboration room is that it captures the changes to the product as the development progresses. Traceability is a key characteristic of PLM and, by the electronic nature of the collaboration room, a history and audit trail of changes can easily be captured and retained. This is an advantage over physical meetings, where the way decisions were made is often lost. Electronic mediation—people using computers to communicate with one another—provides a mechanism for capturing not only the final decision, but the considerations and criteria that went into making the final decision. This information cannot only be used to understand how the particular product evolved the way it did. If organized and made searchable, it can provide other development groups with a knowledge base with which to make future decisions effectively.

Bill of Material and Process Consistency

Bill of Material consistency between functional areas is a continuous and vexing problem. In many organizations, the engineering Bill of Material is different from the Manufacturing Bill of Material, which is different from the financial Bill of Material. This leads to much confusion and wasted time, energy, and material within the organization.

The issue regarding the Engineering Bill of Material and the Manufacturing Bill of Material is best illustrated by the simple part introduced in Chapter 2. This is the part that has a groove down the middle. For engineering, this part consists of a single piece of material and a single operation, grooving the center of the material. However, manufacturing may decide that it does not have the appropriate equipment to perform this grooving operation, so its answer to this problem will be to have three pieces in an assembly operation.

The three pieces will consist of a base and two rails. The assembly operation will consist of taking the two rails and attaching them to the base. The end result is that the finished product is identical in form and function to the product that the engineering people designed, but the material and operations actually used are substantially different.

While the finished product looks the same to both engineering and manufacturing, the implications in terms of material purchased (three pieces of material versus one piece of material), equipment (a milling machine versus an alignment fixture and welding equipment), and manpower requirements (an automated milling process versus manual setup, alignment, and welding) are substantially different between the two manufacturing methods. The labor and material costs, along with the capital equipment costs, will be very different in this simple example. For complicated products, the cost differences can escalate quickly.

If we add to this issue the fact that the financial people may be costing these parts on a different basis from either engineering or manufacturing, then we have an opportunity for the quality of decision making to deteriorate rather rapidly. Engineering thinks the part is being made in a certain fashion with certain characteristics; manufacturing is actually making the part with different characteristics and costing; and the financial people have a third way that does not match either engineering or manufacturing.

Even if the actual Bill of Material is identical between engineering and manufacturing, the Bill of Process may not be. Manufacturing may decide that, based on the mix of products being produced, the Bill of Process defined by engineering is unworkable. Manufacturing may outsource some processes at a higher cost because it is the least costly of the possible alternatives, given the mix and capacity of the plant. However, sales decides that the cost of the product as calculated by engineering means that it can cut the price against the competition, which will increase the volume. This decision compounds the problem for manufacturing, and the organization as a whole is making far less profit than it thinks that it is enjoying.

These situations commonly occur in most organizations and reflect the fact that each of these functional areas uses different systems for their Bill of Materials and Bill of Process. This information is moved from system to system on an ad hoc basis. Product Lifecycle Management has the potential for providing a consistent and cohesive view across these different functional areas. The end result is that all areas are working with the most current and accurate information not only about how the product is designed, but also how it is manufactured, sourced, and costed.

Digital Mock-Up and Prototype Development

Until recently, the only way to really get a sense of the design of a product at scale was to build a mock-up or prototype. Two-dimensional drawings with different views, even if drawn to scale, are inadequate for all the different perspectives, sense of depth, fit and finish of surfaces, and component relationship that a product in three dimensions possesses. While the premier designers and engineers of the past did an extraordinary job of understanding their products with two-dimensional views, even the best of them required mock-ups and prototypes as the product took shape both to understand the complexity of their designs and to discuss those designs with others.

These mock-ups and prototypes allowed product designers and engineers to gather around a real, tangible object; walk around it; look it over; and examine from any possible perspective the design as it would actually take shape. For mock-ups of exterior surfaces, clay was and still is the material of choice for rendering large-scale objects, although plastic is often used for smaller objects and components. The purpose of these models is primarily to examine the visual and aesthetic features of the design, although they are also used to assess surface issues of manufacturability due to surfaces requiring too many or too steep changes in direction.

Prototypes, which are full-scale geometric replicas that may or may not be operable, require craft production since they are literally one-of-a-kind objects. Prototypes are used to assess fit-and-finish of exterior surfaces and other components that must be assembled and fit together. Prototypes are used to assess compatibility of design criteria—for example, looking for fuel lines that are too close to heat sources in automobiles or aircraft, or electronics too near magnetic fields in medical diagnostic equipment. They are used in assessing the visual and manual accessibility of instruments. Prototypes also provide an opportunity for manufacturing engineers to assess manufacturability and assembly issues. In addition, this was the manufacturing engineers' only real opportunity to assess the human ergonomics involved in assembly: getting under the prototype, reaching in to connect one part to another, connecting wiring harness through access points, etc.

The issue with mock-ups and prototypes is that they require moving around atoms. They are time consuming to build. They are expensive. They expend wall clock time and introduce lags into the development process. Although stereolithography is used for some rapid prototyping, a substantial number of prototypes require expensive tooling to be built, which must be discarded if the design changes because of or in spite of the prototype. To the extent that fewer mock-ups and prototypes are built and are built later in the design cycle because information can be substituted for the physical objects, the savings can be enormous. For larger and more complex products, the savings can be in the millions and even tens of millions of dollars.

Thinking lean by trading information for wasted time, energy, and material is the focus of cPD through substituting information for mock-ups and prototypes, reducing the number that are required to be built, and deferring those that do need to be built until far into the development cycle. Because the math-based designs can easily be rendered and manipulated, a good deal of the visual analysis of surface features that mock-ups provided can be replaced by visual renderings. When used with a "power wall," the rendering can pass the Grieves Visual Test. The three-dimensional, life size images are indistinguishable from looking at a mock-up.

Although the relationship is slightly different—rather than walking around the mock-up, the observers are stationary and the representation of the product rotates—the additional advantages that cPD representations can provide over mock-ups are so superior that designers and engineers can easily change from their old practice. Unlike mock-ups, the representations can change lighting perspectives and colors. In addition, different versions can easily be made and compared. Suggestions for improvements by the reviewers can sometimes be implemented in real time to give the reviewers immediate feedback on how those suggestions would affect the overall design.

These capabilities are such that they also replace some of the prototypes' roles. Areas of concern, such as fuel lines and heat sources, can be isolated and color coded to provide immediate visual analysis. Virtual reality goggles and gloves can bring us closer to the Grieves Performance Test by integrating the design and the reviewers' movements to examine ergonomic issues of these designs. Automotive interior suppliers are exploiting this technology today. As we will discuss in the next chapter, the manufacturability and assembly issues requiring the use of prototypes can be addressed by the integration of cPD and Digital Manufacturing.

Digital mock-up and prototype development will reduce the requirement for building physical mock-ups and prototypes, with a resulting savings in time, energy, and material and better functionality. In addition, when it comes time to actually build a mock-up or prototype, the math-based information can directly create the physical model, for example by driving the milling machines that shape clay models. As the images continue to improve—power walls today, virtual reality rooms and holographic images tomorrow—these math-based product representations will allow more and more expensive, time-consuming mock-ups and prototypes to be replaced. Meeting the Grieves Visual Test is effectively here today, and meeting the Grieves Performance Test is not that far off on the trajectory of today's technology in cPD.

Design for the Environment (DfE)

A topic that garners a great deal of attention from senior management is Design for the Environment (DfE). While some organizations have been interested in this topic for a while, most, if not all, organizations have a keen appreciation for the direction that government regulation is moving in, which will in turn drive their interest in DfE. As we discussed earlier, that direction is clearly one of making organizations responsible for the disposal of their products.

From our perspective, DfE has two primary objectives. The first objective is to deal with disposal and recycling in the design process. This entails designing products that have as many components that are recyclable as possible. It includes identifying the recyclable components from other components. It also includes developing the disassembly process and methods of recycling the recyclable components and the disposal methods for the remaining components, especially toxic or hazardous materials. The second objective is to examine the manufacturing process for producing the product and substitute methods, power sources, chemicals, and solvents that are environmentally unfriendly with ones that are less so.

While changing production methods can be accomplished by traditional methods, it is difficult to understand how DfE can succeed in the actual recycling and disposal phase without PLM. While some information about the product, such as product geometry, is embedded in the product itself, the recyclability and disposability of the various components of the product are not. The solution of stenciling or stamping icons on each component to indicate how to disassemble and dispose of the product has applicability only to simple products. While some categories of product, such as batteries, are easily recognized and have known disposal methods, for most categories, without doing expensive tests on the material, there is no intrinsic way for the disposer of the product to know which product is recyclable, which material is disposable normally, and which material is toxic and needs special handling.

If the information regarding recycling and disposability is not extractable from the product itself, it means that this information will need to be acquired from another source. In addition, the longer the period of time for the product's lifecycle, the more likely that traditional sources of information such as product manuals, design specifications, and design notes will be long lost and unavailable to the disposer of the product. PLM with its digital Information Mirroring Model makes it likely that no matter how long the product's lifecycle, the disposal and recycling information linked to the physical product will be available. The product disposer should have available visual representations that are color coded as to the recyclability, such as green for recyclable materials, yellow for disposable materials, and red for hazmat or toxic materials that require special disposal methods.

DfE also impacts the design phase of a product's life. EU directives such as End of Life Vehicle (ELV) and Waste Electrical and Electronic Equipment (WEEE) require that certain substances be restricted to a particular percentage of the overall product. Because substance information is not available, that restriction or allowance is apportioned to each component of the product on the same basis, usually weight. This is a very suboptimal solution, as some components need to use more of the restricted substance, or some components that are homogeneous, such as steel body frames, will use none of the substance. If restricted substances could be apportioned on the basis of the component requirement and aggregated for the entire product, it might be much more efficient than an arbitrary apportionment applied consistently across all components.

The solution that is closely associated with PLM is to create a Bill of Substance (BoS), which would break each discrete component into its substances. These substances could then be aggregated across the entire Bill of Materials so that restricted substances could be apportioned in other than an arbitrary manner (such as percentage by weight) and still meet the overall requirement of the final product. There are a number of issues that need to be resolved, such as organizations not wishing to reveal the composition of proprietary products—such as plastics and resins. However, the trend is toward developing Bills of Substance. This will extend the decomposition of the Bill of Material up one level, as regulations regarding the substances of products proliferate.

Virtual Testing and Validation

PLM is developing a whole range of new opportunities for virtual testing using math-based product information. Organizations can use information about structure and composition to use computers to simulate conditions under which the product is tested. Two examples of the testing that is currently done now are wind tunnel testing and crash testing. In both of these cases, this would correspond to the Grieves Performance Test in that an observer would not be able to discern from instruments and measurements the testing that was done with physical products from the testing that was done in simulation using the math-based product information.

Automotive companies are now using simulation to replace the classical wind tunnel test, where a turbine would be spun up to high rates of speed to simulate wind passing by the object. Since air is not visible to the naked eye, there generally needs to be some sort of smoke that is deployed in the wind tunnel so that the flow of the air can be observed.

In wind tunnel simulation, the air flows are also simulated over the object and, unlike physical wind tunnels, can be changed and modified to come from any angle, or, in some cases, from multiple angles. In addition, it is much easier for the human eye to observe simulated wind tunnel tests because the air flows can be made visible at regular intervals or at certain junctures so that all flows passing over the object can be seen. This gives the tester much more control over the aspects of design that he or she wishes to observe.

In addition, according to an automotive executive,[2] the simulated wind tunnel tests are truer than the actual wind tunnel tests.

This is because the wind tunnel in the physical tests has characteristics, such as being in an enclosed area, that are not present when the wind tunnel tests are simulated. This means that a more accurate view of the wind tunnel results is gained from simulation than from an actual wind tunnel.

The auto industry has also pioneered the idea of crash testing in simulation. The expenses and time necessary to actually crash test a vehicle and the equipment necessary to observe all the characteristics that need to be observed are very high. Simulated crash tests can be performed, and their results almost perfectly matched to the observed results from actually doing a crash test of the vehicle. In addition, individual components can be singled out for analysis and the actual crash testing slowed down in simulated space to determine where the sensitive points that need to be strengthened are. In addition, crash tests at different angles and with different type occupants can be simulated much more easily than they can be done using real cars and real crash dummies.

These are but two of the testing activities that are currently being performed with PLM-based information. Chip designers have done simulated testing, as have aircraft and ship builders. Product testing will be a rapidly proliferating area of PLM. With virtual testing will come better products. Trading atoms for bits means that a wider range of testing with more variations can be done at a fraction of the cost of physical testing.

Virtual testing will continue to accelerate for two reasons. First, as pointed out in the Chapter 1, Moore's Law and its corollaries will increase the virtual testing that is possible. Second, virtual testing professionals will continue to build richer and truer simulated environments. Of all the areas of growth in PLM, we expect the area of virtual testing to be one of the fastest growing. As we will see in the next chapter, the simulation technology of product testing nicely reinforces the rise of Digital Manufacturing.

Marketing Collateral

Given PLM's conceptual basis in promoting cross-functional information usage, it is fitting to have as our final example of cPD the use of PLM in marketing collateral development. For entre-preneurial-driven development, the task of the marketing function is to attract the attention of buyers for the product. This includes not only developing what is commonly called *marketing collateral*, such as product descriptions, fact sheets, and ad copy, but also the material that accompanies the product, such as user manuals and quick-start brochures, and the packaging art work and copy.

While some of this material is artistic and conceptual, a sub-stantial amount must be absolutely factual. The description of the product, its specifications, capabilities, and functions must accu-rately reflect the product. Failure to do so opens up the product manufacturer to charges of misrepresentation and even fraud. In the case of packaging, failure to represent accurately what the package contains can be a violation of federal, state, and local laws, especially in the areas of product safety, food, or drugs.

Without PLM, marketing collateral material suffers from the same problem as all other cross-functional information exchanges. The information about the product is incomplete, corrupt, or may need to be re-created in its entirety. This can range from a simple but embarrassing problem, such as a computer manufacturer's specification sheet stating that the computer has three USB ports when it only has two, to a regulatory violation when a product box fails to list the ingredients of its contents accurately. Procter & Gamble uses PLM to ensure that the latter does not happen. As a by-product, Procter & Gamble uses PLM also to insure that the artwork for all their packages remains consistent and appropriate to the product lines.

The benefit of using PLM for this cross-functional application is at least twofold. Costs are decreased because marketing does not have to re-create information that already exists about product specifications and functions. General Electric has saved a substan-tial amount of the costs of their product manual development by driving it from their PLM system compared with the stand-alone development effort this replaced.

The second benefit is in the increased accuracy of the market-ing material. Since the marketing material is being driven by the product information, it reflects the specifications and functionality of the product in its current form. It is not only engineers who waste time, energy, and material working with the wrong version of the product information. Marketing staffs suffer from the same waste. PLM, with its focus on singularity of information, can eliminate this waste. PLM exhibits the benefits of lean thinking across functional boundaries.

Summary

In this chapter, we discuss how we can use PLM at the beginning of the product lifecycle. The focus on PLM in this stage of the product lifecycle is commonly referred to as Collaborative Product Development (cPD). It focuses attention on the shared understanding that needs to be built as the product design morphs into a tangible product.

The areas of interest that we have focused on in this chapter include mapping requirements to specifications, part numbering, engineering vaulting, product reuse, start and smart parts, engineering change management, collaboration rooms, Bill of Material and Process consistency, digital mock-up and prototyping development, design for the environment, virtual testing and validation, and even marketing collateral. We would expect other areas to develop as PLM matures.

Notes

1. We will use the established practice of a lowercase "c" designation for "collaboration," as in cPD.
2. Presentation by Terry Kline, VP IS&S, General Motors Corporation at the 2004 University of Michigan AUTO^e Conference, April 28, Troy, Michigan.

Digital Manufacturing— PLM in the Factory

Manufacturing may be the function that has the greatest benefit from the application of PLM technology. The manufacturing function is well defined. The object is to produce a product with precisely defined specifications and tolerances utilizing the least amount of resources. There are three distinct phases: producing the first product, ramp up, and producing the rest of the products. Digital Manufacturing is the subset of PLM that focuses on producing a product that consistently meets the specifications, yet does so using the minimum possible resources from the first product on. This chapter will focus on how Digital Manufacturing shows the promise of accomplishing this.

What Is Digital Manufacturing?

The first obvious questions are "What is Digital Manufacturing (DM)?" and "How does it relate to PLM?" Digital Manufacturing can be defined as:

> Digital Manufacturing is an approach involving people, process/
> practice, and technology that uses **PLM** information to plan, engi-
> neer, and build the first instance of a product; ramp that product up
> for volume production; and produce, monitor, and capture for other
> aspects of the lifecycle the remaining instances of that product's
> production using the minimum amount of resources possible.

As in defining PLM itself, we use the word *approach* to indicate that Digital Manufacturing is much, much more than any computer technology. Digital Manufacturing includes the same elements as PLM: people, process/practice, and technology. However, in Digital Manufacturing, there may be more focus on the technology aspect because computer technology is what puts the "digital" into Digital Manufacturing.

The focus on technology also differentiates Digital Manufacturing from manufacturing techniques that also attempt to minimize the resources used in manufacturing but are more people focused, such as Lean Manufacturing. In addition, computer technology is more visible because the manufacturing function is much more heavily weighted to processes that can be captured using computer technology. Because manufacturing is so well defined—its goal is to produce exactly the specified product using exactly the same amount of material, energy, and people's time over and over again—the manufacturing function seeks to turn art and practices into fully defined processes and mirror those processes in computer technology for analysis and optimization.

People also figure differently in Digital Manufacturing. While they are very much a part of the approach in building the first product, their involvement can vary considerably in building the rest of them. In some manufacturing processes, people may figure very little, with machines and robotics performing a substantial portion of the actual production. In "lights out" factories, people may be left out of the approach completely.

In other manufacturing operations, people may perform critical processes. In fact, they may perform them without really being able to define all the parameters they consider. Some casting and grinding operations fall into this category. In its present form, Digital Manufacturing is better suited to more automated or better defined manufacturing processes because it is difficult to use information as a trade-off for time, energy, and material where the parameters cannot be well defined.

In addition, people may be critical in defining the suitability of the output in other functions. This is not usually the case with manufacturing. In the design function, people socially construct what constitutes a good design. In the engineering function, engineers discuss and argue about what constitutes the appropriate engineering implementation of particular functions. At the end of the assembly line, there is not much discussion of what constitutes an acceptable product. It either meets the specifications or it does not. If there is any discussion, it is because the inspection technology is not up to the task of matching specification to actual results rather than because the manufacturing function is trying to socially construct what is acceptable.

Digital Manufacturing is involved in the planning and engineering tasks necessary to build the product to its specifications. In some cases, the planning and engineering tasks are only concerned with specifying the step-by-step process through an existing set of manufacturing equipment, making sure that all the design points are "consumed" in the most efficient way. In other cases, additional manufacturing planning and engineering tasks are required to specify and build the manufacturing equipment, and even the entire factory, in order to produce the product. The latter case is obviously much more complicated than the former. The information requirements for building equipment and factories are also substantially greater than those of the product itself. Estimates are that it requires a hundredfold increase in data requirements![1]

Once the product is planned and engineered, and the first one built, the rest of the products have to be built. Since the first product rarely meets the specifications, tolerances, and timings as defined by product engineering, there is a period of ramp up as the manufacturing group work to remedy the issues. Once the product is produced as desired, the remaining products must be built exactly the same. Since manufacturing equipment changes over time, both from deterioration and the maintenance required to repair that deterioration, the products must be tested at various stages to ensure that they continue to meet the specifications within the tolerances specified.

The final function, monitoring, is capturing the information about the product's manufacture for future use in the product's lifecycle. This is a function that has often been done casually, if at all. We all have wondered when we opened a box with a widget we ordered just who Inspector 32 was and what he or she actually inspected. That probably was the extent of any monitoring for that particular widget, and once we tossed away that little slip all evidence of any monitoring was lost.

However, as we saw in Chapter 4, the trend in government regulations is requiring that we monitor and keep information about each product we produce. Monitoring and recording test information, component serial numbers, and software flash revision levels, if not currently required to be kept and produced, certainly will be required in the future. In addition, this information can and should be used in other parts of the product's lifecycle, such as the service and disposal stages.

Where Digital Manufacturing shows its greatest promise is in "using the minimum resources possible." Digital Manufacturing makes use of information as a trade-off for time, energy, and material. Through simulation of all aspects of a product's manufacture, Digital Manufacturing moves bits instead of atoms.

However, Digital Manufacturing cannot do this unless it is part of the bigger framework of PLM and can work with the information and digital representations developed by other functions, such as design and engineering. In addition, Digital Manufacturing can cycle back to design and engineering its information about the manufacturability of a product. This should reduce the errors, the time to correct errors that do crop up, and prevent designs that cannot be manufactured.

It is this integration of information across functional areas that differentiates Digital Manufacturing from previous point solutions that performed process planning, shop floor scheduling, or some simulation of tools or even entire manufacturing areas. In the spirit of Computer Integrated Manufacturing (CIM), which failed because it approached the problem of cross-functionality from a single-system approach, Digital Manufacturing focuses on integrating the information from different applications with its own systems.

Digital Manufacturing taps into the Info Core and draws out the product information from design and engineering and both feeds back information to help refine and validate the product design and feeds forward information about how the product was manufactured to aid in its use and disposal. Digital Manufacturing taps into the digital substructure pipe of product information and does not require a new "pipe" each time a new product is introduced for manufacturing. While a distinct entity on its own, Digital Manufacturing requires the framework of PLM in order to be successful.

Early Promise of Digital Manufacturing

Because manufacturing is so deterministic, it is somewhat surprising that Digital Manufacturing has triggered the intense interest that it has. The deterministic aspect of manufacturing means that it has already received a great deal of attention through the years. The output, a product with certain definitive specifications, is well defined. The inputs—material, machinery, people, and energy— are also well defined. Thus the processes necessary to produce the output while minimizing the inputs are amenable to analysis and optimization.

The famous industrial engineer, Fredrick W. Taylor, was working successfully on these analysis and optimization problems, or what Taylor referred to as the "scientific method," at the turn of the twentieth century.[2] Taylor's time and motion studies were a precursor to developing processes and equipment that implemented this substitution of information for wasted time, energy, and material. His improvements were dramatic. At Bethlehem Steel, he cut in half the material handling costs by analyzing and computing the optimal size of shovels.[3]

While work continued throughout the century to reduce wasted time, energy, and material in the manufacturing area, there was plenty of waste to work with. The introduction of Lean Manufacturing in the early 1990s found a ready audience because it claimed it could eliminate the majority of non-value-added costs.[4] Even for companies that professed to embrace and implement Lean Manufacturing philosophies, the opportunity to reduce their costs by 50 percent or more still existed.[5]

In spite of a century's worth of focus and effort on reducing inefficiency and waste in the manufacturing function, it appears that there are still major opportunities. CIMdata, a noted research firm in the area of PLM, has done some preliminary investigation of improvements in their manufacturing areas by companies that resulted from their actual experiences with Digital Manufacturing. The results, which are listed in Table 7.1, are spectacular to say the least!

There were improvements in all stages of manufacturing: in making the first product (better understanding of requirements— 50 percent improvement, shortened manufacturing planning process—40 percent improvement); in the ramp-up stage (reduction in number of design changes—65 percent improvement), and in making the rest of the products (reduction in number of work stations—40 percent improvement, more optimized material flow—35 percent). Equally important, there were significant improvements in one of the scarcest resources: wall clock time (reduction in time-to-market—30 percent).

Table 7.1   Digital Manufacturing Benefits Achieved (Based on CIMdata Research)

Benefit Achieved	Percentage Improvement
Reduction in search time for data	80%
Reduction in number of design changes	65%
Better understanding of requirements	50%
Shortened manufacturing planning process	40%
Reduction in number of workstations	40%
More optimized material flow	35%
Reduction in time-to-market	30%
Improved labor utilization	30%
Reduction in tool design	30%
Improvements from better plant layout	25%
More quickly identify areas for improvement	15%
Improved validation of processes	15%
Increase in collaboration/communication	15%
Increase in production throughput	15%
Overall reduction in product cost	13%
Decrease in product design time	10%
Reduction in inventory	10%

These improvement percentages need to be viewed with a degree of caution and skepticism. This was not a study in scientific terms. The sample size was small. Some of the categories were fuzzy and difficult to measure. For example, just how do you measure an improvement in reduction in search time? The numbers were self-reported by the groups that had made the decision to implement Digital Manufacturing. There is no validation that the categories measure what they purport to measure. In short, it is difficult to argue that these improvement percentages are representative of gains that companies could expect from adopting Digital Manufacturing.

In an industry workshop on Digital Manufacturing, workshop participants used percentages similar to those in Table 7.1 in an exercise to quantify the contribution of Digital Manufacturing. The workshop developed a profile of a realistic automotive supplier and worked out what the Digital Manufacturing improvements would mean if applied "conservatively." They worked through the calculations and came up with a substantial number.

Unfortunately, the number was too substantial. As the author pointed out to the group, while the calculations were all correct and had indeed been decreased to be "conservative," the resulting cost savings for this hypothetical company would be double or triple the net profit percentage of even the best-performing companies in this segment of the business. That is not believable. It would be difficult to convince the board of directors looking at the IT Value Map that there is a decrease in costs in one area that would double or triple the earnings of the entire company!

The issue is not that Digital Manufacturing fails to deliver improvements. Early indications are that it does. In specific situations, it may even deliver percentage improvements in the ranges indicated in Table 7.1. As a concrete example of this, a Chrysler automotive plant reduced the number of weld guns from 180 to 17 using Digital Manufacturing simulation techniques.

The problem is that it is inappropriate methodology to extrapolate these improvements to the entire organization. As we will discuss later in the book, in addition to the questions about these particular percentages and what they measure, there are a number of reasons—including the timing of expense reductions and price reductions of the finished product that decrease revenues—why Digital Manufacturing improvements will not translate into income improvements equal to the current cost levels multiplied by the percentage improvements aggregated across the entire manufacturing function.

What trying to extrapolate Digital Manufacturing improvements to the entire organization does do is detract from the fact that Digital Manufacturing shows early promise of substantial improvements in selected situations. All of the areas in Table 7.1 are substantial cost centers in organizations. Improvements of even a few percentage points in areas such as the number of design changes, the number of workstations needed, the amount of labor required, or the amount of inventory needed on hand will financially justify Digital Manufacturing projects.

While indications are that there are situations with tremendous improvement potential, Digital Manufacturing can and is being justified with much more modest improvement targets. Even at the more modest improvement targets, Digital Manufacturing project returns are such that their ROIs or ROAs put them at the top of the entire corporate investment allocation lists.

Manufacturing the First One, Ramp Up, and Manufacturing the Rest of Them

The manufacturing function can be simplified into the following stages: building the first product, ramping up production, and then building the rest of the products. While this is a useful way of looking at the stages, it needs to be pointed out that the reality is a lot messier than this. Rarely is the "first" product the final version of what the first product should be. Ramp up is not always a well-defined stage. It may only be determined that ramp up reached its conclusion at a particular point in time by looking back at production history. There also may be a number of ramp-up phases as product changes occur, and different components of the product may be at various levels of ramp up. Finally, making the rest of them might not be the automatic stage it should be, as manufacturing improvements are still being analyzed and implemented.

However messy reality may be, these different stages are useful to describe the various functions that are performed and the impact that Digital Manufacturing technology can have on these various functions. In addition, both manufacturing professionals and those not that familiar with the manufacturing process can relate to this categorization of manufacturing stages.

Manufacturing the First One

Manufacturing the first product is the first stage of the manufacturing process. Since it is on the engineering–manufacturing border, it is fuzzy where engineering leaves off and manufacturing begins. At this intersection, the manufacturing of the first product is also a large potential source of wasted time, energy, and material as the product "as-designed" is reconciled with the product "as-built." In addition, as our "leaky pipes" diagram in Chapter 3 illustrates, there can be a great deal of ad hoc effort in sharing information between engineering and manufacturing.

Organizations have myriad ways of dealing with this intersection between engineering and manufacturing. Some organizations have manufacturing engineers in the engineering section and attempt to hand off fully specified product designs and manufacturing processes to manufacturing. Other organizations simply hand off the product specifications to manufacturing. The manufacturing function is completely responsible for defining the processes that actually manufacture the product.

Some organizations form ad hoc engineering–manufacturing teams to deal with the hand off. Other organizations have well defined exit–entry specifications that fully specify the exit specifications that engineering will provide and the entry specifications that manufacturing requires. In a significant number of organizations, the exit and entry specifications do not match up.

One organization that shall remain nameless to protect the guilty has a fully operational process that employs manufacturing engineers in both engineering and manufacturing. The manufacturing engineers in engineering fully define the processes, tooling, and routings of the product to be manufactured. The manufacturing engineers in the manufacturing department take these fully developed specifications, decide what is useful to them, ignore the rest of the specifications, and develop their own processes, tooling, and routing plans. Is this a process? Yes. Does it work? Yes. Is it a major source of wasted time, energy, and material? You bet!

If the digital substructure of PLM information is available, then Digital Manufacturing can tap into it in a variety of different areas to replace time, energy, and material with information. It can use virtual space to eliminate wasting valuable real resources. And it can save on the one resource that can never be recaptured, wall clock time. We will look at the various areas where Digital Manufacturing can be employed at this stage below.

Process Planning and Reuse

CAD specifications without process specifications are useful only for generating pretty pictures to hang on the wall. The step-by-step specification of how to actually produce the geometric specifications using physical material and tools is critical to actually producing a viable product. What may look to be a perfectly valid rendition of a product may not be able to be built because the area to be welded or bolted is inaccessible to the available tooling.

There are few manufacturers that continue to use the earlier method of manufacturing process development, which was to design the product, throw it over the wall to manufacturing, and see what came back. Most product manufacturers deal with the process of building the product in the design and engineering stage. However, there are generally two distinct groups: the product designers and the process designers. Since all the current CAD applications have the ability to link process sheets to product design specifications, these two groups do have a symbiotic relationship, if only because of this common linkage.

However, even when this works well and product and process designers coordinate their actions to produce a consistent set of graphical and process specifications, the process plans have to stand the test of the factory floor. The specified tools might not be available; routing might bottleneck other production; or the material and tooling that manufacturing has access to may not correspond to what the planners envision. Or, as is not uncommon, manufacturing might simply have experience or history in building things in a certain fashion that causes it to ignore the process sheets and build products in its own way.

Digital Manufacturing provides value in this area in at least two ways. First, process engineers can use the math-based CAD descriptions of the product and the machines to determine if the process is feasible. Digital Manufacturing tools can determine if the tooling is capable of performing the required operation. Digital Manufacturing tools can check for collisions of the machine and tooling in the product being operated on. Rather than specifying operations that theoretically work, process engineers can validate the operations they are specifying with the exact product and exact machine, tools, and fixtures that are to perform those operations, all in digital form.

Second, the use of Digital Manufacturing integrates the process planning from design and engineering into the manufacturing area and makes manufacturing a partner in the process, rather than a recipient of the outcome of the design and engineering stage. By integrating manufacturing into this process, changes that manufacturing engineering makes to the processes will not only be used as the manufacturing process, but be fed back to the design and engineering function.

The next logical action is to reconcile manufacturing's changes with the design team's proposed process. Not to do so perpetuates the waste of time, both labor and wall clock, as the design and engineering group develops one set of processes and the manufacturing group develops another. This is the equivalent of the company that duplicates the manufacturing process design in both areas.

Not only will this reconciliation lead to an understanding by both design and engineering and manufacturing of the realistic processes needed to actually produce the product, but it will lead to another source of recovering wasted time, energy, and material: reusable processes. Reusable processes are the manufacturing equivalent of reusable parts in the design and engineering function.

With reusable processes that have been reconciled between the design and engineering function and the manufacturing function, the development of processes can begin with a search of a library of processes rather than simply building processes from scratch. In addition, since the reused processes have been tested in manufacturing, some of the time, material, and energy saved in not reduplicating this effort can be dedicated to improving the processes. With the proper categorization and search system, design and engineering planners can produce process plans in a fraction of the time that will be acceptable to the manufacturing group. Digital Manufacturing makes this possible by connecting manufacturing into the PLM information substructure and facilitating the exchange and reconciliation of processes.

Machine, Tool, and Fixture Development and Process

One of the time-consuming aspects of manufacturing the first product is determining exactly how the product must be processed by specific machines and tools. There are two different cases of this. The first case is when the machines and tools are given. The second case is when the machines and tools are to be designed and built as part of the manufacturing process.

In a number of respects, the first case may be easier to deal with, given that the constraints are imposed by virtue of the existing machines, tools, and fixtures. The part or product specification is given. The machine, tool, and fixture specifications are given. The task is to take these constraints and determine a permutation or combination of operations that will yield the specified part or product from the specified machine, tool, fixtures, and raw material.

For simple products or components, this may not be that difficult a task. Manufacturing planners have long performed this exercise. They did this usually by developing an understanding of the simple different tasks they could perform with their machines and then stringing those tasks together to produce the desired result. For more complex components and products, the procedure was to create and make simple structures and to assemble those simple structures into a more complex one.

However, even for simple products, the permutations and combinations of the machines and potential operations multiply quickly. As a result, only a subset of the different permutations and combinations is considered, and once a combination produces the desired result and is within the budget deemed reasonable, the search is stopped.

Human beings are not equipped to do an exhaustive search of all possible combinations or permutations. At best, they can only do so for those components or products that require a small number of operations. They can keep only a small number in their minds, and moving to pencil and paper slows the process.

Computers are under no such limitation. Under Digital Manufacturing, all possible permutations and combinations of operations can be tried. For all but the most complex products, each potential combination can be analyzed for its ability to meet a number of different criteria, such as elapsed time, cost given different machine costs, or material routing distances, and, depending on the decision criteria, the most efficient or least costly solution can be selected. This saves the wasted time, energy, and material that might be employed by a set of operations that is selected because it is the first one that produces the desired component but certainly may not be the most efficient.

When no permutation or combination of operations on the existing machine, tools, and fixtures produces the specified component or product, then there are only two courses of action: change the product or process specification or acquire new machines, tools, or fixtures. Because of the long lead time and the major expense, acquiring new machines, tools, or fixtures is usually out of the question. The product specification is the element that changes.

Before Digital Manufacturing, the process of getting a manufacturable component or product consisted of iterations of designing and engineering product and process specifications: sending them to manufacturing, which tried to manufacture the product according to those product specifications or processes, failed, and sent the problems back to design and engineering, which started the cycle all over again. Digital Manufacturing does these iterations virtually, simulating the processes, determining where there are problems, and correcting them so that time, energy, and material is saved. The cost of moving bits is significantly less than moving atoms.

The second case of producing product when the machine, tools, and fixtures are not given and are also to be built is simpler in one respect, but more complicated in another. The simpler aspect is that by designing the machines, tools, and fixtures we can be assured that the component or product can be produced—at least theoretically. This brings us to the more complicated aspect:

designing the machines, tools, and fixtures is as challenging if not more challenging than designing the product itself. The same gap with respect to the designed or theoretical functionality versus actual functionality has to be closed.

In the past, this gap was closed by building the machines, tools, and fixtures and then enduring a period of shakedown and remediation as the new equipment was tested and the issues and problems from those tests iteratively solved, until the new equipment met the specifications and also actually produced the desired result, often not one and the same requirement. Compounding this process is that this process had to wait on the design of the product, and often the product prototype, before the new equipment could be tested and made to work properly. This often was—and, for those not using Digital Manufacturing, still is—an expensive, time-consuming process that wastes prodigious amounts of time (both worker hours and wall clock), energy, and material.

The move to Digital Manufacturing saves substantial amounts of that wasted time, energy, and material. First, in the same fashion that the product design function benefits from the use of virtual techniques, Digital Manufacturing can obtain like benefits. Machines, tools, and fixtures can all be designed and modeled in math-based CAD systems. While the Grieves Visual Test is useful, the real benefit occurs when the Grieves Performance Test is met and the output of the virtual machines in virtual space is indistinguishable from what would occur with physical machines in physical space.

If the machines, tools, and fixtures meet the Grieves Performance Test, then this new equipment can be fully tested virtually and adjusted as necessary before any new equipment is actually built. In addition, doing this virtually means that there is also no necessity to wait until the product designs are fully done and prototypes made. This method uses the math-based product information to test the functionality and utility of the new equipment. This means that machine, tool, and fixture development can occur concurrently with the product design itself. Again, that is an example of trading information for wasted time, material, and energy.

While the Digital Manufacturing tools and methodologies are not yet fully formed in this area, they may not be that far away. General Motors will shortly require that every bid for new equipment be accompanied by a digital representation of the equipment itself. This digital representation will be incorporated into General Motor's simulated factory to see that the new equipment performs correctly and in concert with the other factory equipment.[6]

Robotics and PLC Simulation and Programming

Robotics and equipment controlled by programmable logic controllers (PLCs) are a unique subset of machines and tools that deserve special attention. These machines and controls perform their operations under the control of a computer rather than a human operator. Because this is the case, robotics and PLC-controlled equipment can bring us a step closer to realizing the Information Mirroring (IM) Model described in Chapter 3 (see Figure 3.3).

Without Digital Manufacturing, robotics and PLC-controlled equipment are subject to the same issues of shake-out and testing. Programmers write programs to perform the operations required to produce the product. The programmers must then run their programs to debug them, as the programs never do what the programmers think they are going to do. This takes time, energy, and material as the machines are busy producing scrap until the programmers get the machines to perform as needed.

Using Digital Manufacturing, robotics and PLC-controlled equipment differ from the equipment discussed in the previous section. With the operator-controlled equipment, the real world–virtual world relationship exists. Virtual equivalents of physical machinery are built in the virtual world. Their operations are simulated and the math-based designs are modified until the required performance is obtained. However, at that point there needs to be human intervention in order to create the physical product in the real world. The physical equipment needs to be constructed according to the math-based design. The resulting physical equipment then has to be tested to see if it produces the same results as the virtual machine does.

Because human involvement is necessary to translate the math-based designs into physical equipment, there is opportunity for error. Dimensions can be misinterpreted, tolerances might not be understood, and human engineers often make "improvements" to the operator controls without understanding the unintended consequences of these changes. In addition, there might be discrepancies between what seems possible in virtual space and what is possible using physical material.

However, where Digital Manufacturing is applied to robotics and PLC-controlled machines the situation is different. The link between the virtual world and the physical does not require human intervention. Assuming that the physical equipment will execute its instructions accurately, which, given the maturity of PLCs and robotics, is not an unreasonable assumption, once the virtual machine's programming code is debugged and the virtual machine produces the desired result, that programming code is transmitted to the robot or PLC. No human intervention with its potential for inducing errors is required.

The opportunity with Digital Manufacturing is that the coding, debugging, and testing can all take place in virtual space. Only when the simulations produce the desired results is the physical machine brought into the loop. The scrap produced by the testing and debugging is all virtual. No time is wasted on the factory floor. Moving bits rather than atoms saves substantial wasted time, energy, and material needed to get computer-based equipment to produce the first product correctly every time.

Ergonomics

There are some factories that run "lights out." In the last half of the twentieth century, robotic machines and computer-controlled equipment moved en masse onto the factory floor. Despite this, the human worker is and will be an integral part of the modern factory for the foreseeable future.[7] Digital Manufacturing must incorporate human ergonomics and factors in order to be able to accurately and completely simulate factory production.

Figures 7.1 and 7.2 show examples of human form simulation as the human forms perform manufacturing or assembly tasks, as depicted by Digital Manufacturing applications from two of the more prominent software providers. Although human beings are substantially more flexible and adaptable than machines, they also have limits in range of motion, angles of approach, weight transfer trajectories, etc. As shown in these examples, the human figure is given the specific requirements of the task to be performed and the simulation is run to see if that task is feasible.

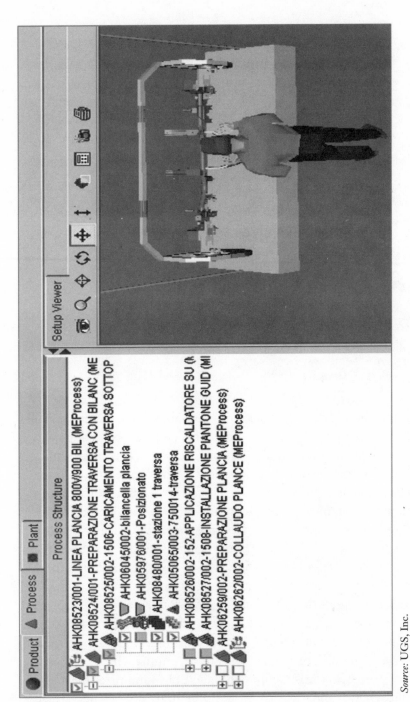

Source: UGS, Inc.

Figure 7.1 Human Form Simulation: Manufacturing Task.

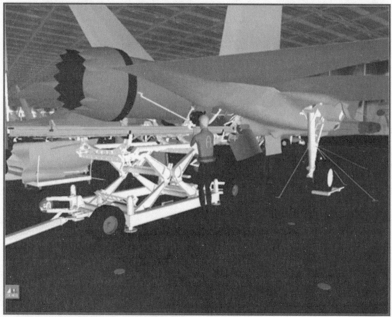

Source: Dassault Systemes.

Figure 7.2 Human Form Simulation: Assembly Task.

If the task requires the wrist to be bent at an angle that is incompatible with human structure, then the task can be redesigned. There are plenty of examples in manufacturing where access holes were cut by enterprising manufacturing workers to facilitate a manufacturing or assembly operation. These access holes showed up on none of the engineering drawings, and, in fact, were not known to exist by the designers and engineering staff. While this may have not only facilitated the manufacturing or assembly, but been the only feasible means of successfully performing the operation, these new holes invalidated the finite element analysis and testing and introduced a potential failure into the product.

However, there are usually many feasible operations that produce the same result. Digital Manufacturing can rapidly perform analysis of the different operations to determine which combination of potentially feasible human operations minimizes the waste of time, energy, and material. In addition, the analysis is performed in the context of the manufacturing environment, so equipment and material placement can also be modified to eliminate the wasted effort of unnecessary steps, movements, and motions. The timing and analysis of alternatives that can be done without a single worker taking to the factory floor would have delighted and amazed Fredrick Taylor!

Ergonomic analysis has to determine not only whether the operation is feasible, but whether it is feasible for a range of human sizes. If the operation is feasible only for a certain size human being, it must be flagged so a decision can be made to incorporate that requirement into the manufacturing specifications or look for an alternate method that is feasible for all ranges of human workers.

In addition, it is not enough that the operation be feasible. It must also be consistent with safe standards that allow human beings to perform that operation over and over again with low risk of injury. Here is a literal implementation of healthy practices. While "best practices" for a particular operation may be an operation that can be performed by 200-pound males in excellent shape, healthy practices are those operations that can be safely performed by all types of human beings in normal condition. Digital Manufacturing's use of simulated activity, while implementing healthy practices, ensures that operations are efficient and safe for a wide range of human workers.

Factory Flow Simulation

The next unit of analysis in the quest for analysis and optimization is the factory itself. Machines, tools, robotics, PLC-controlled equipment, and human beings all come together on the factory floor. While individual operations can be designed, analyzed, and continually improved, how these operations are sequenced has a huge impact on the overall efficiency of production. The goal is not to perform individual operations optimally. It is to perform the entire production cycle optimally.

Factory flow simulation is not something new.[8] What is new is the granularity and the deterministically way in which the factory flow can be done. Previous simulation methods used probabilistic methods to predict flow performance at the factory. This was useful in predicting throughput and bottlenecks. When problems were identified, only the probability distribution and sequencing could be modified. There was no ability to drill down into the individual components so as to modify their behavior in order to ascertain any change in the performance of the factory.

Figure 7.3 is an example of a simulated factory scene. Not only can the entire factory be visualized and simulated production runs performed, but individual areas can be zoomed into and the performance of individual cells or equipment can be analyzed. The factory of the future will be fully specified, simulated, and optimized before the first shovel of dirt at groundbreaking is ever moved. Moving bits instead of atoms will save a substantial amount of wasted time, energy, and material.

Factory flow simulation is not limited to assessing production in a single factory. One issue for major manufacturers is determining which of their plants is best suited for building their new product. What these manufacturers are doing is creating Bill of Systems (BoS) that define the complete capabilities of the equipment of each of their factories. When they need to build a new product, they can match the requirements necessary to build that product with the BoS of their plants. They then can select the plant that best matches the requirements.

Ramp Up

Digital Manufacturing's goal is to eliminate this stage of the product introduction process in real space. This is an idealistic goal

Figure 7.3 Factory Layout Simulation.

because the fidelity of virtual space is such that there will always be gaps between the simulation in virtual space and what actually occurs in real space. In spite of this limitation, Digital Manufacturing has the promise of greatly reducing the ramp-up time of product manufacturing through simulation.

Digital Manufacturing has the potential to reduce a substantial amount of wasted time, energy, and material incurred during the ramp-up process. In the first chapter, we discussed learning curves and the reduction of product costs as production volumes doubles. While learning curves are beneficial, what they indicate is that there is a substantial amount of product that is produced less efficiently than it is possible to do.

The reality is that the ramp-up period has become a luxury that most manufacturers cannot afford. When cycle times were long, the ramp-up period was a small portion of the total manufacturing cycle. However, as we discussed previously, one of the main drivers of PLM is the rapidly decreasing cycle time. As cycle time decreases, ramp-up time increases as a percentage of the overall manufacturing cycle. The waste during ramp-up time becomes more prominent, and the focus is on eliminating or greatly reducing it.

The Virtual Learning Curve

Figure 7.4 shows the experience or learning curve from Chapter 1. However, this time a figure that represents the amount of wasted time, energy, and material is superimposed on the graph. This roughly triangular figure represents the cost wasted during the ramp-up process. What Digital Manufacturing strives to do is save this wasted time, energy, and material by starting virtual production at the top of the experience or learning curve and only starting physical production when the learning curve begins to bottom out.

The ability to actually do this will depend on a variety of factors. For production processes that are highly automated with production operations that are well defined and suitable for simulation, the potential to drive down the experience or learning curve virtually is high. Once the simulations show the required results, the instructions to produce them can be downloaded into the computers controlling the actual equipment.

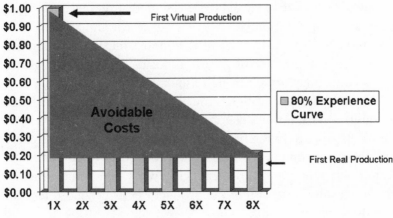

Figure 7.4 Experience Curves

For those processes that have a substantial amount of human interaction, the potential for eliminating the costs of ramp up will be less. While factors external to the human participants can be improved (e.g., reducing steps necessary to retrieve material, rearranging the sequence of actions to eliminate wasted motion), the actual actions of the human beings have to be learned and improved through repetition.

ECO Simulation and Implementation

The challenge of ramp up is more than simply driving down the experience or learning curve. Manufacturing ramp up of new products is invariably accompanied by product changes, commonly referred to as ECOs (Engineering Change Orders). ECOs occur for a variety of reasons. Sometimes the issue is that the product does not have all the functionality that the designers and engineers envisioned. There may be unintended functionality that was unanticipated.[9] There may be issues with suppliers that mandate the replacement of one component with another that requires different operations and processes.

Whatever the issue, ECOs disrupt the drive down the experience or learning curve. In some instances, these ECOs can introduce inefficiencies into the production process that increase costs to a point higher than when production was initially started, because the initial production method has to be unlearned and a new method learned. Production machinery has to be taken down, repro-grammed, and debugged again. The experience or learning curve then has to be driven down from this new, higher point.

PLM as an entire approach in general and Digital Manufactur-ing specifically should reduce the amount of ECOs required. As noted above by CIMdata, companies utilizing PLM have reported an astounding 65 percent reduction in ECOs. Increased ability in the design and engineering function to determine if the product design actually delivers the desired functionality reduces the need for ECOs.

As PLM tools and technologies continue to develop, designers and engineers will have a better ability to meet the Grieves Perfor-mance Test. This means that designers and engineers can deter-mine that their product designs actually function as expected, not just look like the intended design. This should reduce the instances of functionality failing to materialize or unintended functionality cropping up.

Digital Manufacturing, by providing better coordination, com-munication, and simulation in the first stage of manufacturing, will further decrease the need for ECOs. Problems with supplier com-ponents, operations, and even product functionality can be sur-faced and corrected without these problems ever making it to the manufacturing floor.

Finally, for the ECOs that are unavoidable, Digital Manufactur-ing can reduce the disruption of introducing the ECOs into the manufacturing process. The ECO introduction can be simulated to determine the least disruptive method for installation. Automated equipment can be reprogrammed and debugged in virtual space. New work instructions can be designed and workers trained off line. Only when the issues surrounding the ECOs have been fully worked out virtually will they be moved to the factory floor. While there may be temporary disruption and increase in inefficiency, use of Digital Manufacturing will reduce the impact to a minimal level.

Manufacturing the Rest

While Digital Manufacturing plays a critical role in manufacturing the first product and attempts to eliminate manufacturing ramp up, it needs to continue to play an instrumental role in manufacturing the rest of the products. Digital Manufacturing applications are the source of information that can be used to replace wasted time, energy, and material as products continue to flow through the factory. Digital Manufacturing has information about how the products should be manufactured. It can use that information both to compare against the actual production of products and to capture information about the actual production for use in other stages of the product's lifecycle.

Areas that Digital Manufacturing can play a part in are planning the production in the factories, assisting in the actual production by providing production information as required, monitoring and auditing the production process against the specifications, and capturing exactly how the product was built for future use.

Production Planning

Production planning is an obvious use of the information developed to create the factory flow simulation and analysis. Rather than attempting to allocate equipment time and schedule production on the basis of aggregate numbers (e.g., there are X hours of available milling machine time, Y hours of stamping time, and Z hours of welding machine time), the use of factory flow information allows production planners to use discrete timing and flow information to efficiently allocate production and simulate that production to determine unnecessary equipment changeovers, bottlenecks, or other production flow problems.

In a great number of situations, production planning is simply finding a schedule that works. Little attempt is made to reduce the waste of time, energy, and material by finding better schedules, even if finding the best or optimal schedule is not feasible. In addition, production schedulers generally need to build in slack time, because their information about processes and operations is incomplete. This further adds to the inefficiency.

By using the models developed for factory flow simulation and analysis, production planners can develop production plans and test those plans in simulation. Because they have detailed information about operations and processes, they can schedule production more tightly because they do not have to incorporate the slack time required when this information is unavailable. They can also better handle changes in production schedules, which cause huge problems for organizations not using Digital Manufacturing. At its best, the ERP system is tightly coupled into this production planning process. Order information drives the production planning process, and the scheduling information for both customers and vendors is fed back into the ERP system in a cooperative fashion.

Computer-Aided On-Demand Builds

Digital Manufacturing can do more than simply be a repository of information. As shown in the Information Mirroring Model, the information in virtual space can be provided in real space wherever and whenever it is needed. One of the reasons companies make large batch runs is that the information requirements necessary to deal with different products is overwhelming for human beings. While human workers are extremely flexible and adaptable, they do not have the precise memory recall necessary for manufacturing diverse products. When faced with the requirement of making a diverse set of products, human beings need memory assists like Bills of Materials, work notes, and blueprints.

Even then, their error rate for selecting the correct components increases, because recall from viewing the information and remembering it as they go to pull that exact sequence of parts is imperfect. Through time and repetition, the error rate decreases and the amount of different product variations in the workers' repertoire increases.

This may have been acceptable 100 years ago, but it is unacceptable today. The time required to search for, retrieve, and review the proper Bill of Materials, work notes, and/or blueprints between products is simply unavailable in today's production facility. The error rate as workers learn various product configurations is also unacceptable. The first product has to be built flawlessly the first time. Finally, because of decreasing cycle times and increasing complexity, the ability of workers to build a repertoire of products is greatly diminished. The worker who built a stable product for his or her entire working life is a thing of the past. Modern workers will be required to build different products on shorter and shorter cycles.

One way to cope with this information overload is to build in batches. This does not eliminate the problem and creates new, more serious problems of its own, such as inventory accumulation. In addition, there may still be some raggedness in performance as the product is changed over and a new batch starts. The other solution is to provide the worker with the information about the product to be built as he or she needs it. In addition, the information can be provided in a form that cues the worker in the most direct and appropriate fashion.

While terminals can provide work notes, Bill of Materials lists, and even color-coded blueprints, other communicating mechanisms, such as lights, can provide information in a focused fashion. Nissan builds dozens of different dashboards on the same assembly line by lighting up the appropriate parts bins in the correct sequence at each station. This cues workers which dashboard among the dozens possible is to be built. Assembly workers need only to focus on how the pieces go together, not which pieces they should select.

The product has already been built in virtual space. It is simply a matter of getting that information into real space in a form that eliminates wasted time, energy, and material. While experiments with terminals, sensors, lights, RFID chips, and other information-conveying mechanisms will continue, Digital Manufacturing already can make an impact on the inefficiencies of the batch method of production by facilitating the efficient building of different products.

Quality Control Monitoring and Audit

In virtual space, every product is built exactly the same way each and every time. The information about each instance of production is indistinguishable from any other instance. Assuming no hardware malfunctions or software bugs, an expectation we rely on with great confidence, quality control is never an issue in virtual space. Specifications will be met each and every time.

Real space is not so accommodating. Moving atoms around is a much more iffy proposition than moving bits. Machines do not perform exactly the same operations to the same degree of precision each time. Material is not as uniform as virtual space assumes it to be. And human beings vary in reliability and competence in performing specific operations.

It is why we need to spend a great deal of time and effort in devising and implementing inspecting mechanisms and regimes. At one time, the thinking was to produce the complete product, inspect it, and rework the product as necessary to bring it into specification compliance. The wasted time, energy, and material of this model and the result that the product was inferior in quality convinced manufacturers that quality needed to be something that a product possessed all along, not something added at the end of the manufacturing cycle.

Using Digital Manufacturing to ensure the quality all along is one of its natural functions. Digital Manufacturing has the information specifying the product in virtual space. Its task is to compare that information with the information about the product in real space and determine if a gap exists. If the gap is not within tolerances, the Digital Manufacturing system takes appropriate action: notification, reporting, intervention. This means that the inspection equipment must be able to communicate its results to virtual space as shown in the Information Mirroring Model. Only if it does that can the comparison be made.

However, it is quickly becoming insufficient simply to inspect and verify a product's quality. It is rapidly becoming a requirement to provide proof of that quality. Rapidly coming into force are regulations regarding safety that require the manufacturer to capture this information and provide an audit trail for independent inspection. If an issue arises with the product at a later date, then this audit trail will provide verification that the problem was not in its manufacturing.

This information is also invaluable in the increasingly litigious legal environment. A seat manufacturer for the automotive industry captures information from its manufacturing equipment on the torque force for bolts installing the seats into cars. It calculated the potential liability for injury when a seat broke lose in an accident and would be required to answer "yes" to the plaintiffs' attorney's question, "Isn't it possible the seat bolts were not tightened to the proper level?" They would be required to answer "yes" because it would theoretically be possible that the seats were not tightened to the proper level. The inability to eliminate this possibility was much more costly than capturing that information as production occurred.

As we will discuss when we look at product liability in the next chapter, ignorance is not bliss. Inspecting ensures that gaps between product specifications and the actual product build are identified. Bringing that information into virtual space improves the process that requires dealing with those gaps because everyone has access to that information. It is not just an inspection report sitting in a paper file somewhere. Capturing that information allows the manufacturer to have confidence that the gaps were dealt with and that the products sold to the consumers were produced to its specifications.

"As-Built" Capture

As human beings, we all come with "as-built" plans. It is called DNA. Unfortunately, products do not have the same information mechanism built into their structures. As a result, when "as-built" is needed later in the product's lifecycle, there are only three things we can do: disassemble and inspect, deduce, or generalize. Disassembling and inspecting is the only definitive method of retrieving this information. When we absolutely have to have the information (in such instances as product recalls or our example about the helicopter part), we have to track down the product, disassemble it, and inspect it.

As noted previously, this is a substantial waste of time, energy, and material, especially if we are only interested in a small subset of all the manufactured product. The use of RFID technology will help this issue, but this still requires being in close physical proximity to the physical product. Being able to interrogate virtual space is tremendously more efficient.

Where there is not a requirement to know with absolute certainty the product makeup, those wanting "as-built" information often resort to deduction. "If the product was manufactured on this date, then based on our purchase and usage of a certain component, the part installed was probably this specific part." The problem, as we have often experienced as consumers bringing our product in for repair, is that the deduction is often faulty.

After opening the product for repair, the service organization finds the part is incorrect and needs to order a different part. What was going to be an efficient process—order part, bring in product, replace part—has turned into a very inefficient one—order part, bring in product, order another part, return first part, replace part. This is a waste of time (both the time to do duplicate work and the time during which the consumer has his or her product out of service), energy, and material.

The third method, generalization, moves an information search requirement onto the user. This is a common method of dealing with service manuals. Rather than specifying which product we have and how we must perform corrective action, the manual gives us general information and then, for those things that differ between models, gives us instructions for dealing with the variations. Even if we have a good understanding of which model we actually have, a major assumption when it comes to technical components, the effort of tracking the correct information can be daunting.

However, since the manufacturers build the product, they can and should know what specific components went into each instance of the product. With the other information that Digital Manufacturing deals with, capturing "as-built" information would seem like a natural function for Digital Manufacturing. Not only is this information needed in the manufacturing stage itself, but it is highly useful for other stages of the product's lifecycle, such as service and disposal.

Summary

As we saw in this chapter, Digital Manufacturing is not only an effort in its own right, it is also an aspect of PLM. It is in its infancy, but shows early promise of reducing wasted time, energy, and material in the manufacturing function. This will minimize the use of resources in the manufacturing function. Digital Manufacturing is working in the three stages of manufacturing: building the first product, ramp up, and building the rest of the products.

Digital Manufacturing drives the interface between design and engineering and manufacturing in making the first product. It facilitates process reuse and simulates the operations of physical equipment. For robotics and PLC-controlled equipment, it takes us one step closer to the Information Mirroring Model and transfers the information in virtual space to computers controlling real space, eliminating a source of human intervention.

Digital Manufacturing is attempting to replace the physical ramp up with a virtual ramp up. In both optimizing the manufacturing process and in dealing with changes to the product, it attempts to perform as much of the learning and experience possible by moving bits instead of atoms.

Finally, Digital Manufacturing uses the information it contains about the product to make the rest of them more efficiently. It helps plan production. It assists in flexible manufacturing. It compares information about what is actually occurring against what should occur to produce quality products. Last, but certainly not least, it captures information about how the product was built that can be used in other aspects of the product's lifecycle.

Notes

1. Private conversations with Information Systems & Service (IS&S) Process Information Officers (PIO) of General Motors Corporation.
2. For a first-hand explanation of his scientific method, see F.W. Taylor, *The Principles of Scientific Management*, New York, London: Harper & Brothers, 1911.
3. See J.R. Beniger, *The Control Revolution: Technological and Economic Origins of the Information Society*, Cambridge, MA: Harvard University Press, 1986, pp. 293-298. Beniger does an excellent job of describing the whole idea of using information to control and thereby make more efficient a wide variety of industrial processes.
4. See J.P. Womack, D.T. Jones, D. Roos, and Massachusetts Institute of Technology, The Machine That Changed the World: Based on the Massachusetts Institute of Technology 5 Million Dollar 5-Year Study on the Future of the Automobile, New York: Rawson Associates, 1990.
5. See J.K. Liker, *The Toyota Way: 14 Management Principles from the World's Greatest Manufacturer*, New York: McGraw-Hill, 2004, for a description of how manufacturing is as much or more a philosophical or belief orientation as it is a methodology. The word *way* has the Eastern connotation for this more so than our word *approach*. See pages 10-12 for the example of the improvement of a supposed lean manufacturer.
6. Private discussion with GM Information Systems and Manufacturing executives.

7. Those who want to believe that the humanoid robots depicted in the 2004 movie "I, Robot" are realizable near term would do well to remember that Isaac Asimov wrote the stories upon which that movie was loosely based in the 1950s. While we have made great strides in computers and robotics, we are nowhere near the robots Asimov envisioned. Interestingly enough, an automobile company, Honda, has been working on humanoid robots. Their robot, ASIMO, is interesting, but no danger to even the most diminished human worker.

8. See I. Stahl, "GPSS: 40 Years of Development," paper presented at the Proceedings of the 33rd conference on winter simulation, Arlington, Virginia, 2001.

9. These are referred to as product "bugs." Less scrupulous manufacturers cast them as additional "features."

Outside the Factory Door

E VEN THOUGH PLM logically begins with the front of the life-cycle, it promises some of its greatest benefits in the later stages, after the product leaves the factory door. Without PLM, information about the product's performance in use and its actual quality is lost to the detriment of other functional areas. In addition, product usability is a problem because of increased product complexity. Functions that are being mandated by regulation, such as recycling and reuse, become cost prohibitive without PLM or a PLM-like approach. This chapter will explore these opportunities and issues.

Costs Do Not Stop at the Factory Door

There are a number of proponents of PLM who believe that the real value of Product Lifecycle Management will come in the later stages of the product's life or, in other words, once the product leaves the factory door. Capturing and using information about the product when the product is in use should be a major opportunity for lean thinking, or trading information for wasted time, energy, and material. This information from the use of the product or, as we shall call it, "in-use" product information should prove valuable both in reducing costs and increasing revenue for the organization.

Although rarely reflected in the unit cost calculation of an organization's accounting or costing systems, the cost of the product does not stop once the product leaves the factory door. There can be substantial costs to the organization that are directly attributable to each product once the product is sold and in the hands of the user. In days past, when the term *caveat emptor* reigned supreme, organizations stopped having any liability for the product once title was transferred to someone else, normally a distributor or reseller. From that point, the cost of operating the product, the cost of maintaining it, the cost of any harm done by the product, and the cost of any disposal or recycling was borne by the user. However, nowadays, due to contractual, legal, or regulatory changes, these costs lie with the product producer and not with the owner of the product.

In this section, we will look at three sources of cost that do not stop at the factory door. These are quality scrap production costs, product liability costs, and warranty costs.

Quality Scrap Production

Quality scrap production costs are those costs associated with products that fail in their use, although they have met the specifications at the time of their manufacture. Currently, because of the lack of integrated product information, a typical situation is that these products fail in the field and are caught in some sort of repair cycle. Simply put, the product fails and the user brings it to a repair facility for repair or replacement. If the product is in its warranty period, as defined by elapsed time from sale, distance traveled, time of usage, or some other metric, the repair facility will repair the product at no charge, usually by replacing a malfunctioning component. In the absence of abnormal usage, the problem is that the failing component did not meet its functional design requirement—it did not meet its requirement of functioning for the required amount of time or usage.

Unless it is a safety issue, a product producer expects a certain percentage of failed products. Even under Design for Six Sigma (DFSS), there is a probability that a component of the product will fail. However, when this statistically expected failure rate is exceeded, the issue may not be random failure but some systemic problem with the product design and specifications. If the failure rates exceed this expected value, engineers get involved to analyze, identify the root cause of the failure, and redesign a solution that eliminates it.

At this stage of the product's lifecycle, the problem is usually not that the component fails to meet its specifications. In today's environment, quality control within the factory has caught and fixed those problems. The issue is that the component does meet its functional requirement. The assumptions that produced the requirements-to-specifications linkage were flawed.

This necessitates going back and reassessing the function-to-specification linkage and its associated assumptions. Because the functional requirements are a given and have been warranted to the product user, the general solution is to change the specifications, which will necessitate a change in product design. The change needs to go through the validation and testing process to ensure that there are no unintended consequences that cause new and different problems.

In the meantime, the manufacturers of these components are chugging away, producing as many new components as their purchase orders call for. It does not matter that they are Lean Manufacturers of the first order. They are producing scrap. It is quality scrap because it meets current specifications, but it is scrap nonetheless.

Some of it cannot be helped. It takes time to collect, analyze, and correct this type of issue. However, that does not explain the lapse of weeks and sometimes months that component manufacturers routinely experience between the time that the issue is understood and a redesign that corrects the issue is finalized and the time that the component manufacturer is notified.

This means that the component manufacturer wastes time, energy, and material in a number of ways, all expensive. Because the component does not do what it is supposed to do, there is a waste of design resources to redesign the component to meet its functional requirements. From the time the problem is identified until the time a replacement component is ready for the manufacturer, the component manufacturer produces defective components that everyone knows will all have to be replaced. During this period, production costs are doubled!

Finally, the component manufacturer has to pay the product manufacturer for the resources to de-install the defective component and reinstall the replacement component. This is a losing situation at best that is then compounded by the elapsed time from when the problem is known to when it is communicated and resolved. This can and does lead to substantial financial exposure.

Some of the functionality-to-specification issues that create this faulty production will be eliminated by PLM earlier in the product lifecycle. As we saw in Chapter 6, the ability to perform digital prototype development, validation, and testing will allow for a much wider range of validation and testing procedures to be performed at a fraction of the cost. However, even with those improvements, there will still be some issues because the virtual world will never completely reproduce the physical world and all its variations. The ability to capture and communicate information quickly and accurately, especially about product failures, will be crucial in reducing the wasted expense of quality scrap production.

Product Liability

The second type of cost that we will look at is product liability cost. This is cost over and above the repair and replacement costs that occur when some harm has ensued because the product has failed to perform in actual use. While money cannot compensate for the injury or death that occurs when a product fails to act as represented, there is no lack of personal injury attorneys ready to assuage the victims of these product failures with as much money as they can recover, minus of course their not inconsiderable percentage.

In fact, even if the product was not the cause of the death or injury, plaintiffs' attorneys are only all too ready to hypothesize the product's involvement if it is even in remote proximity to the injury- or death-causing event. In the past, the strategy of the manufacturers was to operate under a philosophy that ignorance was bliss. Their reasoning was that if information was not available to the plaintiffs, then the plaintiffs could not point to faulty design, improper assembly, or insufficient materials as the cause of the accident.

However, the "ignorance is bliss" strategy often turned out not to work. It more often than not gave a substantial opening for a plaintiff's attorney to ask the representative of the product-manufacturing firm on the stand at trial, "Isn't it possible that the part did not meet specification?" or "Isn't it possible that the bolts were not tightened to the specified level?" Even if the representative were very confident that the manufacturing processes were such that the part met specifications or had been tightened to the appropriate level, he or she could only answer honestly that it was possible. Without the data from the actual assembly, the representative has no way of knowing whether that specific part met that specific specification or was installed in the specific manner called for. Ignorance was not bliss, and there are many manufacturers with multimillion and even multi-hundred-million dollar judgments against them to prove otherwise.

The only sure way to answer the question, "Isn't it possible?" is to have the data from the actual design and installation of each product. Traceability is a key characteristic of PLM, and the traceability that PLM provides is the only real defense as to whether or not the product met its specifications. Ignorance is not bliss, and ignorance about how the product is designed or assembled is management malpractice, along with being a potential legal liability waiting to happen.

This traceability of product information and its visibility does two things. First, if there is an issue in the design or manufacturing process, the time to correct it is when it occurs. If we are collecting information and data as the design and manufacturing process is occurring, then we can produce immediate exception reports to show that one part is not fitting with another part or that bolts are not being tightened to the specified and required level. The implicit justification in the past was that the cost of collecting this information outweighed the product liability costs, but it only takes a few hundred-million-dollar-plus judgments to skew that equation fairly dramatically.

This is one aspect of product information at the design engineering and production stage that is critical in the defense against product liability. However, the second aspect is to collect in-use information about the state and usage of the product so that the cause of the damage can be assessed appropriately. It is not unreasonable to assume that in a number of these cases where the plaintiff's attorney hypothesized improper action on the part of the manufacturer, the true cause was misuse by the user.

In the case of automobiles, speeding, failure to use a seat belt, and sudden steering changes are likely sources of automobile crashes and injuries. Automotive manufacturers are developing black boxes similar to those that are found in airplanes to capture information about the product use at the time of impact. This in-use information about the speed the automobile was going, the braking time, whether seat belts were worn or not is all information about the product that could be used as a defense in product liability cases. This in-use data could be invaluable in reducing product liability costs, not only for automobile manufacturers, but for manufacturers of a wide range of products that are subject to misuse by the user.

Warranty

The final cost concerns that of warranty cost. Warranty costs are a huge source of costs in many organizations. In 2004, warranty costs for U.S. manufacturers were in excess of $20 billion dollars. For Hewlett-Packard, Dell, Ford, and General Motors, warranty costs exceeded a billion dollars in each of those organizations. IBM's warranty costs were under a billion dollars, but only slightly.

In-use product data would be exceptionally useful in reducing the costs of warranty. Many organizations use unrelated third parties to perform warranty repair and reimburse those parties on the basis of the work that has been performed. In some cases, the more expensive the part that fails, the more warranty reimbursement the third-party service organization obtains. Without in-use product information about the failing of that particular product, there will be a temptation for these third-party producers to replace the most expensive part associated with the failure as opposed to the part that actually did fail.

Today, manufacturers spend a great deal of time and money in assessing their warranty repairs and attempting to determine whether the product replaced was the part that failed and assessing what changes they need to make. In the worst case, they can spend money in redesigning and retrofitting parts that are perfectly adequate, but are replaced because of the compensation system in place to the third-party repairers, and never really getting a handle on the problem. In the best case, they are forced to police their third-party repairers and to retest the parts and weed out any parts that are functional but have been replaced because of the compensation system in effect.

In either case, in-use information about the product, its use, and its failures would reduce these warranty costs. Manufacturers would have a clearer idea of what failed and why. They could feed this information back to the product designers, who could begin to determine which functionality to specification linkage was flawed. At a minimum, design engineers could incorporate this information into future designs in order to reduce future warranty costs. For many organizations, it would change warranty from a payment system into a learning system.

Quality in Use, Not in Theory

A major opportunity to collect information about the product in use is in the quality control area. Currently, quality control stops at the factory door and is based on meeting specifications, as opposed to being based on whether the product actually functions in use. The definition of *quality* that is currently in use in our organizations is that the product meets a certain specification. The assumption in use here is that the specification that the product is measured against will give certain performance. However, the true definition of the quality of a product is whether it actually performs in use, not whether it meets the specification. Until the product performs in the context that it is represented to be capable of performing in, there is no assurance that the linkage between specifications and use is well defined.

While there is a tacit attempt to link product specifications with in-use capabilities, the collection of that information is sparse and subject to errors. We can point out one source of information about the product in use—warranty information—that can be flawed because of compensation issues. Surveys regarding product capabilities are biased, and often reflect the perception of the last segment of use of the product, as opposed to its use over a period of time. One way product designers attempt to get around these deficiencies is to test their product, and there is no question that manufacturers do extensive testing of their product. However, these tests, by definition, are fairly limited and staged. The real information on a product's performance is from the actual use of the product. If that information could be collected and analyzed, it would provide a much wider range of data in a much wider variety of environments than testing results, warranty information, or survey information.

One of the major uses of this in-use information is to enrich the database of simulation information so that testing can be done on a simulated basis. Because we have the product information available to use, and as computing power and product lifecycle management systems get richer and more robust, the ability to simulate the environment and the use of that product in the environment will continue to get better and better. In this case, the in-use product information is still valuable because we can use that information to compare against our scenario databases of testing information to ensure that as new uses of the product or new extremes of the product's performance are defined, we can capture that information and change our testing and simulation models so that we can have a better handle on new products that we're developing. As a result, this integrated view of Product Lifecycle Management information and the math-based representation of the information plus the data coming from the use of the real object in real-world circumstances can be combined to create a powerful increase in the quality of the product from today's standards.

As part of the quality issue, the in-use information can create better linkages between specifications and functional use. As mentioned above, today's engineers design their product to specifications with the assumption that those specifications are linked to specific functional uses. However, these linkages are theoretical at best. When information is not available about the actual use of the product, the tendency for engineers is to increase the specifications so as to over-engineer the product as a safety measure. Without in-use part information, there is no way to know what the costs of this over-engineering are. In some cases, we may be able to relax tolerances because the in-use data will show us that the product is not being used to its design specification. With the data available to use, we should be able to do a much better job of designing products to a functional specification rather than to a theoretical specification.

Product Usability

While costs may not stop at the factory door, the expertise to understand and fully exploit a product's functionality may. The increasing functionality being added to products to increase their value and the enormous flexibility that programmable software enables has increased the complexity in configuring and/or operating today's products.

There are two aspects to this issue. The first is ensuring that product documentation, which includes manuals, tutorials, and in-use prompts and messages, reflects the product's actual functionality and related operations. The second aspect is enabling the product users to be able to find and understand the documentation when they need it.

The first aspect was addressed in Chapter 6. Having a single source of product information through PLM that drives product documentation should enable product documentation that is derivable from product design. This differs from the usual process in most organizations today, where there is a completely separate exercise in deciphering the product's functionality and operation after it has been built.

The second aspect, enabling the product users to be able to find and understand the documentation when they need it, is partly an information availability issue and partly a design philosophy issue. If, as has been often said, the first casualty of war is the plan, then the first casualty of a product's use is its documentation. If product manuals ever get unpacked at all, they usually find their way to, at worst, the waste can or, at best, a drawer full of other manuals where it can successfully hide.

With the proliferation of the Internet, some manufacturers make their manuals available on their Web sites. However, finding where that is and decoding the product model numbers to find the right manual can be daunting. As products acquire more computing and communication capability, we would expect to see product documentation, tailored to that exact product, to be embedded in the virtual product and accessible anywhere and anytime.

We would also expect design philosophy to evolve with the availability of in-use information. If design engineers were able to obtain data about what functions of their products were used and how, this might help them develop more useable products. In-use information could tell them what prompts were abandoned, an indication that users did not understand how to proceed to achieve a certain function, or what functions were never used, an indication of function creep that should be eliminated.

PLM is needed to improve product usability of our increasingly complex products. The blinking time on VCRs that was beyond a significant subset of the population to remedy is a thing of the past. However, product usability will be an increasing problem that PLM can help improve.

New Revenue Model Opportunities

In-use data and information will allow product designers and engineers to validate their assumptions about the functionality-to-specification linkage more than ever before. Information about how the product actually worked versus how the product was supposed to work will allow designers and engineers to close that gap. This data and information will feed back to designers and engineers so that they can improve later versions of the product and will feed forward so that new generations of products can be designed and engineered better, with more effective functionality-to-specification linkages.

However, in-use data and information can also be used to create new revenue opportunities. If we examine the Information Mirroring Model introduced earlier, we can exploit the characteristic of PLM that we referred to as *correspondence* to develop revenue-producing opportunities. In fact, some product manufacturers are already doing so.

Repair Services

If the product manufacturer maintains correspondence between the physical product and its virtual equivalent on an as-built, as-maintained basis, then there is a revenue opportunity for that manufacturer to offer repair service contracts at a competitive advantage to other service providers. With this physical-to-virtual correspondence, the manufacturer is in a position to know what specific components are in the product. So if repairs are needed, the disassembly and inspection step can be eliminated. Additionally, this correspondence means that the manufacturer also has a record of the repairs made to the product, so it can avoid wasting time troubleshooting the same problem.

If this physical-to-virtual correspondence is augmented by intelligence within the product, any change in the status or environment of the product can produce an update to its virtual equivalent and trigger an alert to the manufacturer, which can correlate the status changes against normal behavior and spot abnormalities before the advent of full-fledged failures.

This correspondence is at work today with high-ticket items, such as the Joint Strike Fighter (JSF) jet. Under pre-PLM approaches, problematic status changes would cause a trouble light to be illuminated. The pilot would report the problem to his or her crew chief upon landing, who would investigate the problem. Even if the repair crew had a pretty strong suspicion about the problem component, they would have to open the plane and visually identify the specific model in order to be sure they were ordering the correct component. Only after they did all that could they order a replacement component. This procedure wasted time, energy, and material and, most importantly, irreplaceable wall clock time.

When the PLM approach is used for the JSF plane, if the plane's computer detects a problem, it immediately notifies the crew chief and his or her repair group electronically, even if the plane is in the air and on a mission. The repair group can diagnose the problem, and, since they have a virtual image of the plane and its components, they can immediately place an order for the exact component necessary for that specific aircraft. While they still have to open the plane and replace the part when the plane returns to base, they do it in a single, efficient process. They do not have a plane with its parts strewn around while they identify the problem, order the replacement part, and wait for it to be delivered.

It is obvious that this type of capability makes sense for jet fighter planes. However, it also makes sense for a wide variety of products. Automobiles are a logical choice for this capability, since they possess a significant amount of computing capability. As Moore's Law continues its progress, more and more products will have computing intelligence that will enable this capability. Who would not pay a fee to have the repair person show up before the garage door opener or the garbage disposal failed?

Product Extension Services

The business problem with product sales is that they are usually one-time events. The manufacturer makes the product. The end user buys the product. The manufacturer now needs to find a new customer for the next product it makes or wait around for an old product to be consumed or wear out in order to sell to the same customer. The model works for products that are consumed quickly such as razor blades, but does not work so well for durable products that have much longer lives.

The repair services discussed above are a mixed blessing for the product manufacturer. While it is a source of revenue, repairing a product means that the product failed. This is usually not a happy event for the product owner and may serve to cause that owner to search for a different supplier with the hope eternal that the new supplier's product will be more reliable. The exception to this is repair services that anticipate and prevent an actual failure.

The ideal for any organization is a product that has a reoccurring revenue stream attached to it. In some cases, the reoccurring revenue stream is associated with how the product is used. This works in some cases, but not in others. HP is widely reported as making a significant portion of its profits on ink for its printers. The automobile industry does not make a dime on the gasoline its customers' cars use.

As computing capability is built into more and more products, there will be opportunities to create different versions of products and collect in-use information so that customers can be charged for the product based on actual usage or additional capability.[1] It might be possible, for example, to sell a washer with basic capabilities to newlyweds. As the washer detects the increase in usage caused by the addition of children, it offers the opportunity to purchase upgraded capabilities. The same versioning philosophy might come into play by analyzing driving habits and conditions and offering an upgrade to tune performance and save on operating costs.

The ideal is use of the correspondence of the physical to the virtual representations to offer continuous monitoring to detect abnormal changes. General Motors offers this today with their On-Star® system. This service monitors the status of the car and detects a change in status of the vehicle, such as airbags deploying. If we pay a monthly fee for monitoring our cars, might we not pay a monthly fee for monitoring ourselves through medical devices that we may have implanted in us such as pacemakers or insulin pumps?

End of Life Recycling and Disposal

The investment and execution of Design for the Environment (DfE) and Cradle-to-Cradle[2] (C2C) initiatives are wasted expenditures of resources if the information about the product is not available at the time of the product's end of life. As we noted earlier, the time from a product's introduction into service until its end of life can be decades. Due to decreasing cycle times, there may generations upon generations of new products and associated information that has been generated subsequently. Because of the relative newness of PLM, it is unclear how well its applications will function in identifying and retrieving decades-old product information.

Within product producers, there currently is a fair amount of the attitude, "We've done our part in designing our products to be recyclable. If they wind up not being recycled, it's not our concern." There is no doubt that a substantial portion of today's products being designed to be recycled will end up dumped in a landfill. Currently, the focus is on being certain that potentially hazardous or toxic substances are handled differently from ordinary waste and prevented from reaching the landfills. Batteries, transformers, and tires are some of the components that need to be disposed of by authorized facilities. However, it is a certainty that a fair amount of those objects are reaching the landfills.

However, as we noted earlier, the regulatory climate is changing. Sooner or later, and most likely sooner, it will not be enough for product manufacturers simply to design their products for disposal and recycling. Product manufacturers will be responsible for the actual disposal and recycling. The European Union's regulations for automobiles and electronic devices make the product producers responsible for the disposal and recycling. That trend will continue, both in product scope and in additional jurisdictions.

PLM will also drive the trend because, with PLM, compliance will become easier to enforce. As product identification technology becomes more and more prevalent for other uses within PLM, this same mechanism will allow regulators to identify the manufacturer or manufacturers of products that are not complying with their disposal and recycling regulations. Compliance may also require providing documented evidence that recycling and disposal have occurred. Not being caught disposing improperly will not be enough. Companies will have to prove that they have complied with disposal regulations. From a technology perspective, the scenario of garbage trucks being equipped with RFID sensors and not picking up trash that has items that need special recycling and disposal methods is not that absurd.

However, compliance is achieved, the trend is clear. PLM information about disposal and recycling will need to be maintained for long periods of time and made available on demand. If the organization is controlling the recycling and disposal, the requirements will be one of ensuring the product information remains available over potentially long periods of time. If, as will be the case for the majority of products, the disposal and recycling is done by an independent third party, there will need to be work done on making this information available in some common form. In addition, manufacturers will need some notification that their product has been appropriately recycled or disposed

While the trend toward regulation in the area of recycling and disposal is evident, it is unclear how stringent the requirements for product manufacturers will be or how strict compliance will need to be. For those organizations that are committed to these environmental issues, it is not enough simply to design or manufacture their products with DfE techniques and standards. With PLM, they will have the capability to facilitate the actual disposal and recycling at the product's end of life. However, much work needs to be done at this stage to make that happen.

Summary

This chapter focused on the role of PLM after the product leaves the factory door. This is probably the least developed aspect of PLM, but it may, in the long run, produce the greatest benefits. In-use data and information can be fed back to the design cycle to improve new versions of the product, and fed forward to produce new generations of the product.

There are substantial costs that can occur after the product leaves the factory door. Three of these costs are product liability, quality scrap production, and warranty costs. A quality product is not one that meets specifications, but one that performs well in use. PLM can be used to identify and substantially reduce these costs and improve quality. Ignorance is anything but bliss.

Finally, exploiting the correspondence characteristic of PLM can create new revenue opportunities in repair, monitoring, and versioning. PLM allows the product lifecycle to be completed successfully by providing the information necessary to fully recycle and dispose of the product long after it has been produced.

Notes

1. The computing capabilities being integrated with products will open up the opportunity for products to be viewed as software. For an excellent treatment of the economic opportunities here, see C. Shapiro and H.R. Varian, *Information Rules*, Boston: Harvard Business School Press, 1999.
2. See W. McDonough and M. Braungart, *Cradle to Cradle: Remaking the Way We Make Things* (1st ed.), New York: North Point Press, 2002.

Developing a PLM Strategy

AS IS EVIDENT FROM the previous chapters, PLM is an enterprise-wide and, in some cases, a supply-chain wide issue. Because of that, strategies at varying levels and over varying time frames need to be developed and coordinated. This chapter will discuss the issue of strategies and sub-strategies, the organizational levels that need to address them, the metrics that need to be employed in measuring their progress and success, and the results and returns that should be expected.

What Is Strategy?

There is a great deal of confusion that surrounds strategy and its development. Anyone who has had to get three children to three different functions—a baseball practice, a soccer game, and a gymnastics class—that all start at exactly the same time have an intrinsic understanding of the development of strategy and a keen appreciation for its execution. However, when we talk about it in business, there is a great deal of confounding strategy—which, strictly speaking, is a plan—with the goals and objectives that we are seeking to accomplish.

People confuse goals and strategies. "Our five-year strategy is to be market share leaders." In fact, there are some who believe that the goals and objectives are inseparable from the plan to achieve them. Supporting that view is the common usage of "strategic plan" that refers not only to the actual plan to accomplish goals and objectives, but the goals and objectives themselves and also the analysis that goes into assessing the organization's and environment's suitability for such goals and objectives.

In keeping with this common usage, here are the elements that we believe must be part of a strategic plan for PLM: a vision of the future, a realistic assessment of where we are today, a plan of action for bridging the gap between the reality of today and the vision of tomorrow, and the capabilities and resources necessary to carry out that plan.[1] In our view, PLM is not a goal or objective. No one has a goal to implement PLM. Rather, PLM is part of a strategy to reach some goal or objective. PLM is an enabler of the goals of the organization, but is not an end goal in itself.

A Vision of Tomorrow

The most important component of the strategic plan is a vision of tomorrow. If there is no perception of what we think tomorrow will look like, then there really is no need to produce a strategic plan. Even if our vision of tomorrow is a duplicate of today, meaning that we are simply continuing status quo, that is a legitimate vision of tomorrow. Our plan then must consist of determining how to keep what we have rather than to accomplish anything new. But maintaining the status quo is not only a legitimate endeavor for us to undertake, it is an extremely common goal for any number of organizations.

However, without a vision of tomorrow, inertia takes over. Whatever it is that we are doing today, we will keep doing tomorrow, and whatever that brings us, it will bring us. This can appropriately be called the "Que sera, sera" strategy.[2] This is a process orientation rather than a goal orientation. Some companies are successful with it. While they do not have a vision of what tomorrow will be, they know that the process that they have used in the past has brought them acceptable results. Their assumption is, "If

I continue to do this process, I will continue to be pleased by the results." This is generally a dangerous assumption. When conditions change, these organizations find themselves unhappy with the results because they did not have a vision of tomorrow, only a process for today.

Visions of tomorrow for organizations almost always include the organization's being larger and more profitable. As we saw from the IT Value Map, this translates into increased revenues and lower costs. Because of its impact across the entire organization, PLM is a logical part of a strategy to increase revenues and lower costs. PLM decreases costs through Lean Thinking, trading off information for wasted time, energy, and material. The resources freed up can drive an increase in product variety, quantity, functionality, and quality.

One technique in use to develop a vision of tomorrow is to bring the senior leadership together and ask them to imagine themselves, say, five years in the future, then look around and describe what they see. While a lot of their descriptions will simply be extrapolations of the past ("We're bigger"; "We have more locations") there will be some valuable insight into what this new, future environment can hold for the organization ("We've moved all manufacturing to China"; "We have moved from ceramics into fiber optics"). Comparing the new and different assumptions, which explain how the novel view of the future came to exist, with the current assumptions might allow for plans and resource allocations that justify a new approach to the business.

PLM, in particular, benefits from an approach like this because some of the unintended benefits from collecting and sharing information might not be obvious. It is only when we take a new and discontinuous look at the future that we might perceive novel and creative uses of information that are not simply extrapolations of what we have done in the past.

There is one very common vision of tomorrow that drives many PLM initiatives. This vision is that of "One Company." This phrase, with the specific company's name following "One" (e.g., "One General Motors") is a commonly used rationalization for embracing PLM.

The reason that the "One Company" theme resonates so strongly as a vision of tomorrow is that these companies have, as we have discussed earlier, increased dramatically in scale over the past 30 years. Since increases in scale were due not only to organic growth but to growth through acquisitions, these companies now find themselves with a myriad of different organizations and divisions that all operate differently. As a result, these companies suffer from inconsistent performance and an inability to take advantage of their size and scale. They have an inability to share from a common knowledge base, and they incur unnecessary costs of time, material, and energy in order to coordinate the activities of the various operations. Their thinking is anything but lean.

Within the operations themselves, the methods of doing things may be fine and, in point of fact, may even be efficient within that group. However, across the entire organization there are tremendous amounts of inefficiency and waste because of the duplication of effort and the coordination and reconciliation of different processes and practices that perform the same tasks. In addition, for the executives at the top of the organization, the different processes and practices simply drive them crazy. Because all these methods are different, the top management of the organization simply does not have the bandwidth to be able to understand all the varieties of and variations on ways of doing things. As a result, all these organizations lack transparency and, when an issue does arise, top management has to spend valuable time and effort in trying to understand the idiosyncratic nature of how a particular operation performs before management can even begin to address the issue.

This makes it very difficult for, say, the engineering vice president. When deciding which group will design a particular new product, he or she has to structure the definition of that product for the specific processes and practices of the operation that will develop that product. Furthermore, this engineering vice president has an inability to parcel out the work to different operations because he or she has no way of defining how they are going to communicate and coordinate. In point of fact, that coordination and communication will be extremely time consuming for the engineering vice president first to define and second to monitor, as the project progresses.

It is no wonder that, for these kinds of organizations, the "One Company" vision of tomorrow is an imperative corporate initiative. However, in addition to the management view of how it impacts their jobs, there is also a strongly held belief that, if management could get to a "One Company" organization, then there will be two major benefits. First, cost will be reduced by saving the time, energy, and material that is being spent for duplication and coordination efforts. Second, management will be able to increase revenues by increasing the functionality and quality of their product with those resources that are freed up under the "One Company" initiative.

PLM obviously is not only very compatible with this "One Company" vision, but is also a means to implement it by building a substructure of product information that all the operations and their functions, such as engineering, manufacturing, etc., can share. PLM can also implement common processes and practices. PLM technology helps to structure this "One Company" view. In addition, where this product substructure is shared with the supplier community, these suppliers are also structured by being part of these processes and practices and sharing from this common base of product information.

In addition, this "One Company" view is a one-company view, and not a "One Engineering" or "One Manufacturing," or "One Service" view. While these organizations start by bringing the functional areas into common processes and practices, the idea of building an informational core where the different functions can share information in order to substitute this information for wasted time, energy, and material is also a major driver for the organization.

This "One Company" view is a reaction to the pendulum between specialization and coordination swinging to the coordination side. Organizations split up their functions in order to be able to improve their productivity much like Adam Smith's pin makers did, but the coordination efforts decrease some of this productivity. PLM is an attempt to use the technology to reduce the costs and inefficiencies of coordination efforts among areas of specialization.

A Realistic Assessment of Today

However, in order to get to this vision of tomorrow, there needs to be a realistic assessment of today. If we do not know where we really are, it will be difficult to go where we want to go. If we are unrealistic or even delusional about our current situation and capabilities, then we will be starting toward our vision of tomorrow from a faulty starting point. Since the normal tendency is to overestimate our situation and capabilities, the resources and requirements that we will need to reach our vision of tomorrow will be grossly understated.

In the next chapter, we will talk about performing a realistic assessment of where the organization is. But that is only one of two pieces. The other piece is an assessment of where the environment is. As it pertains to the assessment, the environment is made up of customers, competitors, and governments. We need to assess each of them in turn in order to understand how we obtain our vision of tomorrow. While we may be perfectly happy with the pricing structure of the products that we are producing today and completely happy with our processes and practices that do that, our customers and competitors may have different ideas.

Our customers may decide that we need to decrease the price of our products because of pricing pressure they experience with their products. Our competitors may decide that they want our customers, so they reduce the price of products that are equivalent to ours. As an example of this, automotive suppliers are faced with a 5 percent decrease in cost each and every year, whether they like it or not. The question is not whether these automotive suppliers will decrease their prices, the question is, will they keep their customers if they don't. The answer seems to be a resounding "No." In addition, competitors, especially global competitors, are changing the equation in terms of the cost and functionality of products. In order for organizations to continue to sell to the same customer base they must predict and factor into their plans the actions that competitors are likely to take.

Finally, as we saw in a previous chapter, regulation plays a big part in determining organization activities. Complying with regulations is not a voluntary activity. Companies either decide that they will comply with them or they will not be in business. As a result, they have to factor the change of regulations into their plans. If their view of tomorrow shows them selling the product without any concern for how the product is going to be disposed of, it is not a vision of tomorrow. It is a daydream. Organizations need to factor in the regulatory climate as they develop their vision of tomorrow.

Irrespective of what these changes are, organizations will continue to need to Think Lean as they use information to trade off for more wasted time, energy, and material. We may not know exactly what will happen with customers, competitors, and governments. However, what we do know is that if we do not develop and maintain the information about our product, we will not be able to make informed decisions when the environment changes. Information is our only defense against uncertainty. With information, we can at least plot out plausible courses of actions and understand their impact. If we do not have that information, we are at the mercy of the changing environment.

Plan for Bridging the Gap

The plan for bridging the gap has to include the three aspects discussed in Chapter 5: people, processes and practices, and technology. All three of these elements have to come together in a coordinated plan for an organization to get from where it is today to where its vision is for tomorrow. All of these elements are required if an organization is to make this transition. If one of the elements is not addressed, the whole plan suffers. For example, if the right software is acquired, but people are not trained to use that software, then people will not use the software effectively, if at all. Little or no benefit will be obtained. If processes are made more efficient, but these processes cannot be implemented in the software, then these efficiencies will be lost.

We can learn from other implementations that involve people, processes, and technology. According to one study by Gibson[3] on IT risk assessment, leadership, employee perceptions, and project scope and urgency cascaded together to provide a multistage approach to risk assessment. As shown in Figure 9.1, if leadership, employee perception, and project scope and urgency were all positive, companies were able to perform major software implementation and bring up the system for all of its employees simultaneously in what is commonly called a "big bang" approach.

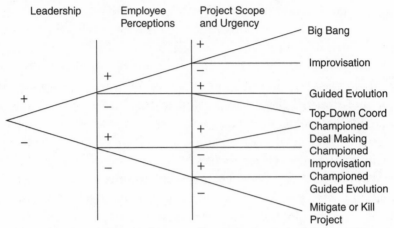

Source: C.F. Gibson, "IT-Enabled Business Change: An Approach to Understanding and Managing Risk," MIS Quarterly Executive, 2(2), 104-115, 2003.

Figure 9.1   IT Risk Assessment

Conversely, if leadership, employee perception, and project scope and urgency were all negative, then the project needed to be mitigated or killed. Projects where there was neither leadership nor the employee perception that the project was favorable and, to compound the problem, where there was a lack of understanding about the project's scope and urgency would not make it past the planning stage. But, as the figure shows, there were variations of how these projects could proceed under different conditions.

One main conclusion we can draw from this research is that if leadership were positive, then that could be the driving force for even an incremental approach to the project implementation. However, if leadership were negative, then it required somebody in the organization to step up in order to be able to champion or negotiate with the rest of the employees to drive at least parts of the project to a successful conclusion.

Capabilities and Resources Required

No matter how lofty the vision or how necessary that vision is to the organization, the reality is that capability and resources either constrain or enable strategies to be undertaken. While an organization may have an excellent vision of what tomorrow should look like, a great assessment of their internal capabilities, and a realistic view of what their competitors, customers, and government are doing, if they don't have the resources and capabilities to carry out the plan, there is little possibility of their realizing their vision of the future.

For all organizations, there is only a finite amount of resources, and those resources are allocated on the basis of perceived value of return to the organization. Eventually it all comes down to the IT Value Map, where decisions to be made about investments revolve around the increase in revenue, decrease in costs, and size of investments in order to make a decision on the viability of the initiative.

That is not to say that there is always a careful calculus of the benefits of a certain strategy and the investment necessary to implement that strategy. Sometimes visions of tomorrow are fuzzy, even if they are the right visions. The leadership of the organization may decide that the risk/reward profile is too attractive to forgo. It is also unnecessarily limiting to move to forecasts and budgets too early in the strategy cycle. Along with being natural killers of joy, accountants sometimes are too quick to develop budgets and make decisions about what they know today. This does not necessarily make them wrong, but it is only one of the possibilities. The future is rarely an extrapolation of the past. Investments in new technologies, such as PLM, often open up new opportunities that were unforeseen at the time the investments were made.

In PLM, the danger with respect to resource allocation is the Goldilocks Effect, asking for too much or for too little. Because PLM can have such a large reach and impact within the organization, there can be a tendency to develop a comprehensive vision of tomorrow and a desire to allocate the entire resources to enable the plan to accomplish this. The problem is that the larger the request, the more detailed the plan and the greater confidence in the outcomes the decision makers responsible for resource allocation will want.

Because PLM is still in its initial stages and evolving, the detailed plan required to support an allocation of the size necessary to create and implement an organization-wide vision of PLM might never be completed. Revising the plan as the technology evolves and practices change to use that technology can easily be a never-ending process. The problem is that decision makers like stability and concreteness, at least when they are allocating huge sums of money, and constant revisions give them the impression that the fund requesters do not have a concrete and stable plan.

On the other hand, requesting too small a resource allocation means that PLM could be viewed as peripheral and unimportant. PLM is not an approach that can be done on the cheap. There can be a tendency to focus on some of the practice aspects of PLM, such as rich communications, and decide that "a chat room for engineers" means literally that, and that simply installing Instant Messaging (IM) will implement PLM. Even though we have been careful to label PLM as an approach and not overemphasize the technology, technology is a critical component of PLM. Creating virtual spaces that enable information singularity, correspondence, cohesion, traceability, reflectivity, and cued availability cannot be done with IM.

The right approach to PLM requires developing the vision of the future, even if it is a little fuzzy, but it also requires developing clearer way stations along that path. "Chunking" up PLM into manageable and well-defined pieces so that decision makers have confidence in the plan and its outcomes is imperative. Meeting those outcomes before approving the next tranche of resources allocation allows the decision makers to gain experience and confidence in the capability of PLM on other than blind faith in the vision of the advocates of PLM.

While resources are in finite supply, with the right vision and plan, organizations can find internally or acquire externally resources for initiatives that create value as identified by the IT Value Map. If PLM can impact both costs and revenues across the organization, it should compare well with other initiatives and be a priority to fund. Even for organizations that have limited internal resources, capital markets, bankers, and venture capitalists are all in the business of evaluating and funding initiatives, if they can be convinced that value can be created. Organizations should be able to provide that justification for PLM.

Resources are needed to enable the plan for PLM, but capabilities are required to execute the plan for PLM. We discussed the capabilities that people in the organization need in order to deal successfully with the change that PLM brings. However, we looked at this from the perspective of the recipients of the decision to implement PLM. From the perspective of driving this change, the capability that is required to create the vision of the future, devise the plan to get there, and then make that plan happen is leadership.

As we will see later, leadership is a critical factor in determining whether an IT initiative will be successful or not. While there are many, many aspects of leadership that pertain to an innovative and complex approach such as PLM, there are three aspects that we would like to highlight. The first aspect is the role of the leader in melding change and technology. The second aspect is the requirement that the leader adapt his or her view of the future and the plan to get there as conditions change and evolve. The third is the leader's role in communicating the strategy throughout the organization

Change is difficult enough, but coupling it with introduction of new technology amplifies the problem. The change and technology requirements of PLM are of a nature that organizational leadership cannot simply approve the plan and sit back waiting for status reports. Changing the nature of functional boundaries that have served as an organizing cornerstone is difficult work that requires constant attention. Breaking new ground in developing a substructure product information pipeline that becomes a new major asset of the organization is groundbreaking work. Since "best practices" are a myth, leaders must spend time in determining how to tailor healthy practices and ideal practices to their own organization.

Since PLM is an evolving area, the vision of the future will also surely evolve. Given the external drivers of changing scale, scope, complexity, cycle time, and regulatory pressures, the probability that the organization's leaders will have 20/20 vision of the future is remote if not completely nonexistent. Strong and active leadership is needed in order to evolve the vision of the future. Leadership is needed both to adjust the plan to changing conditions and to capitalize on the new opportunities that building an informational core for their products will bring.[4]

Communicating the planned and emerging strategy may be the most critical component of leadership in this situation. Managing may be all that is necessary when we want people to continue doing what they have done in the past. However, when we need people to change we need leadership.

People need clear communication not only about what changes are required of them, but also about why those changes are required. Coercion is usually not an effective tactic in eliciting change. The most effective tactic to accomplish change is to have people want to change. That is best accomplished by being able to have people see the same vision of the future that the leadership sees. The better leadership is able to communicate that vision of the future and its ensuing benefits for the organization, the easier it is for people to accept it, even if it takes them out of their zone of comfort.

A vision of the future and the plan to realize it are simply elaborate daydreams if the resources and capabilities are not there. Resources are constrained in any organization. However, the right vision of the future clearly articulated and evaluated for creating value should allow organizations to allocate the needed resources to PLM. Leadership is the key capability required in executing the plan and making the vision of the future a reality. Adapting the vision enabled by PLM and its associated plan to a rapidly changing environment, driving the change necessary to realize that plan, and communicating it effectively are critical components of leadership that deserve our focus.

Impact of Strategy

What is the impact on the organization of developing a strategy for PLM? Why not just begin installing PLM applications, or even entire systems, to implement the individual areas we have talked about previously, such as vaulting, engineering change management, factory simulation, warranty reporting and analysis? Each of these areas can be justified on the basis of individual ROAs or ROIs. Why go to the trouble and expense of having to create a comprehensive vision of the future? It is difficult, contentious work to develop and build consensus on a vision of the future.

While individual PLM projects can and should be justified on the basis of their own ROAs or ROIs, we would argue that there are major differences in the sizes of the ROA and ROI returns when they are coordinated within an overall strategy. Individual PLM projects can be incompatible, and the ensuing selection of different technologies will require expending time and additional resources at some point in time to standardize software, repurchase software, and write-off the old software; retrain people; and reevaluate processes and practices. Any time there is "re-anything" in the sentence, it means resources have been wasted.

The question of the impact of PLM strategy has been examined by PLM research firms in order to attempt to quantify the effect of developing a PLM strategy. As we can see from Figure 9.2, which is taken from an AMR research report, the hypothesis is that PLM payback schedules vary both in magnitude and in time frame, depending on the scope of the PLM initiative. What AMR proposes is that short-term PLM projects have a 1-to-1 payback, mid-term projects have a 10-to-1 payback, and longer-term projects have 100-to-1 payback. While AMR presents little evidence to support changes in magnitude depending on the scope of the PLM initiative, it is very plausible that the level of returns of PLM initiatives increases as the scope moves from short-term projects to long-term strategies.

As the AMR chart shows, individual projects have a payback of six to nine months, and are focused on automating and standardizing collaboration within the functional area. Vaulting, part numbering, and collaboration rooms are examples of functionally based PLM projects. In the mid-term, AMR proposes that operating

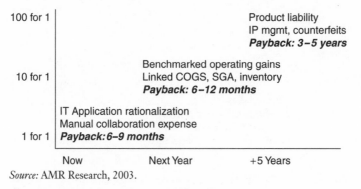

Source: AMR Research, 2003.

Figure 9.2   PLM Payback Schedule

gains are linked to cost of goods sold, SG&A, and inventory. For these initiatives, AMR proposes paybacks of 6 to 12 months. These initiatives reflect returns that are accrued from sharing information across functional areas, such as engineering to manufacturing or manufacturing to sales and service.

The real payoff that AMR proposes is the development and execution of a PLM strategy. In this situation, they predict a 100-to-1 payback of investment as a result of a strategy for Product Lifecycle Management. In the chart, AMR mentions as examples of strategic goals the management of an organization's intellectual property, counteracting product counterfeiting, and preventing product liability. These are all areas that can cost organizations huge amounts of money that would be impacted by a coherent PLM strategy executed across the organization.

We would like to clarify AMR's chart with respect to the type of PLM initiative under discussion and propose the PLM payback schedule in Figure 9.3. Short-term PLM returns are garnered by implementing applications, primarily within functional areas. Mid-term returns are obtained by implementing systems that cross two functional areas. The long-term, higher payback is generated by producing and implementing an overall strategy for Product Life-cycle Management that impacts the entire organization and extends into the supply chain.

One of the reasons that there is substantial payback to strategy is that developing strategy does not require a huge consumption of

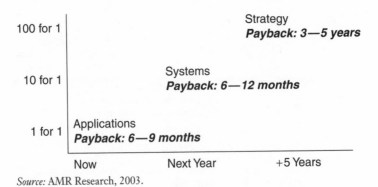

Source: AMR Research, 2003.

Figure 9.3 PLM Payback Schedule—Applications, Systems, Strategies

resources. While it does take a concerted effort on the part of an organization's leadership to create a shared vision of the future and devise a plan to get there, it would seem difficult to find a higher priority activity. Without such a vision and plan, the organization is a reactive entity, being buffeted by continual winds of change. Lean Thinking requires that the leadership make the best use of its resources. It does that by planning and executing a strategy that drives it in a clear direction to its vision of the future.

Implementing a PLM Strategy—Lessons Learned

Once a decision is made to approve and initiate a PLM project, the question then becomes, "How can it be made successful?" While PLM is a unique undertaking, organizations have engaged in cross-functional systems in the past. We should look to those kinds of projects to see what we can learn. One such cross-functional system is obviously Enterprise Resource Planning (ERP).

ERP in its various forms has been around for more than a decade. Thus, we can look to these systems in order to see what separated successful implementations from unsuccessful ones. One such study by Brown and Vessey[5] looked at ERP systems in order to determine what factors differentiated successful ERP implementations from unsuccessful ones.

Table 9.1 lists the five success factors that Brown and Vessey isolated from companies that were successful in ERP projects versus

Table 9.1 Five Success Factors for ERP Projects

- Top management is engaged, not just involved
- Project leaders are veterans and team members are decision makers
- Third parties fill gaps in expertise and transfer knowledge
- Change management goes hand in hand with project management
- A satisficing mindset prevails

Source: Brown, C., and Vessey, I. (2003). "Managing the Next Wave of Enterprise Systems: Leveraging Lessons from ERP." *MIS Quarterly Executive*, 2(1), pp. 65-77.

those companies that were not successful. Those success factors are (a) top management is engaged, not just involved; (b) project leaders are veterans and team members are decision makers; (c) third parties fill gaps in expertise and transfer knowledge; (d) change management goes hand in hand with project management; and (e) a satisficing mindset prevails. We will look at each of these factors and see if and how they apply to PLM

Top Management Is Engaged, Not Just Involved

Engagement and involvement are two very different things. There are numerous committees that senior management, directors, vice presidents, and even CEOs sit on within their organizations. In a number of situations, top management is on these committees to give legitimacy to the initiative and not actually be involved with it. In a number of cases, top management never actually attends any of the meetings of the committee, but simply sends its delegates, who represent the executives on that committee.

This may be acceptable for certain types of projects where an air of legitimacy is all that is necessary for the committee to perform successfully. But for Product Lifecycle Management initiatives, real decisions about cross-functional allocation of resources and responsibilities have to be made. As noted before, within the functional silo approach to information, decisions about expending resources that may appear rational and responsible for the specific functional area (such as engineering or manufacturing) may be suboptimal or even wasteful when looked at across the entire enterprise, or even across the supply chain. Engineering may decide that it is not worthwhile, based on its priorities and resources, for it to collect information that manufacturing engineering needs in order to produce the product or that sales and service needs in order to be able to support the product.

All the functional areas are under resource constraints, and it is difficult to find an organization that would not have this perspective when faced with this situation. It is perfectly rational and logical for engineering to say, "I will cut out information that is needed somewhere else so that my operation can save these resources and operate as efficiently as I need it to for my purposes." Manufacturing and sales and service's response can only be that they will re-create the information that they need at the point and time they need it. However, as demonstrated earlier, this can be a substantial waste of time, energy, and material when looked at from the perspective of the entire organization.

The problem is that issues of this sort have to be resolved at the top of the organization where the functional areas meet, because resource allocation takes place at this level and not lower down within the organizations. Design engineers and manufacturing engineers with working knowledge of the problem may have to surface the issue of re-creating information that can be captured with less cost elsewhere. However, the final decision on how to reallocate resources so that a minimal amount of resources is expended for the task from the perspective of the entire organization will have to be made by top management.

If top management is not engaged in making these decisions, what generally happens is that the status quo prevails and no change occurs. In addition, in the absence of top management's engagement in these decisions, the natural tendency will be to make a decision that favors the resource use for the department or individual function itself, as opposed to benefiting some other department. It is human nature for people to attempt to benefit the group that they know rather than the one that they do not know.

Unfortunately, the expert advice on all corporate initiatives is that top management needs to be engaged and not involved. Internal controls, human resources, customer resource management— all of these initiatives have as one of their main principles that the engagement of top management is a necessity. While everyone argues that his or her initiative is the most important one and should have top management's involvement, obviously top management has only a limited amount of time and attention to go around.

The two arguments that could be used to bolster the position of PLM versus all these other initiatives are these: First, PLM is about creating a valuable asset of the organization, namely intellectual property concerning the products of the organization. This is in addition to showing real cost savings and revenue opportunities. A number of these other initiatives are about cost savings or even about revenue opportunities, but they do not at the end of the day create a sustaining new asset. PLM does.

The second argument is that, given that information can be more beneficial to the overall organization than it may be to one department, decisions about allocating resources to create information where it is the most logical and efficient to be created need to be made by top management. In a number of these other initiatives, the purpose of top management involvement is to ensure that execution of the plan is proceeding, as opposed to deciding where resources need to be expended for cross-functional purposes.

Project Leaders Are Veterans and Team Members Are Decision Makers

In many of these project committees, the committees themselves are stocked with junior members of the department, with people who have time on their hands, or with individuals who are dispensable within their specific department or function. This is not a surprising phenomenon, because most organizations do not have their best people sitting around waiting for something to do. It is a difficult proposition to go to the most valuable members of our group and say, "By the way, in addition to the critical responsibilities that you currently have, we also need you to undertake this major project."

However, with PLM, it is critical that the members of the team have both deep knowledge and understanding of their departments and functions and the ability to make decisions for their organization without having to go back and seek permission from other managers. The issue of deep knowledge is an important one. Experienced members of the various functional areas will also have an awareness of aspects and issues of other departments and functions and not be solely focused on their own issues.

It is difficult, but not impossible, for someone to spend a great deal of time within his or her organization and not be exposed to other issues. There are always cross-functional issues, and the more experienced an individual is within his or her department or function, the more likely it is that he or she will have been exposed to these issues. As a result, more experienced individuals will recognize the value of performing some additional work in their area that will benefit another area. They understand this because they will eventually be called on to solve some problem that this lack of information will cause for another department or function. Veterans of the organization will understand those issues and be more amenable to allocating resources to solve them once and for all, as opposed to taking the myopic view of the issue that a more junior person might take.

The veteran managers also will not be as susceptible to falling into the trap of agreeing to what appears to be a small change that has major ramifications for their area. When junior members are on the team, they will often agree to what appear to be minor changes in their processes or practices only to take them back and find out that they have major ramifications. They then have to go back to the committee and renege on those commitments.

Because this is a sheer waste of time and causes the erosion of trust in that team member, this is an event that ought not to occur. The most dysfunctional of all situations is when some of the team members are veterans and some are junior personnel. The veteran personnel will either become disgusted with the committee or view it as a colossal waste of time and act appropriately, or they will attempt to game the junior members in order to benefit their own functional area at the expense of the entire organization.

It is also imperative that team members be decision makers. PLM is not a matter of implementing the status quo. It requires fundamental changes in how the organization views its processes and practices. If team members are not decision makers, valuable time and effort will be lost because team members will have to go back to their respective organizations and attempt to explain why reallocation of resources is necessary. Again, the status quo is a very powerful source of inertia, and in a great number of cases the team members who come back trying to reallocate resources will be met by the belief that the status quo is good enough.

For PLM to be successful within the organization, the team members must work out the right allocation of resources, make the decisions within the committee room, and then be able to execute on those decisions and implement them. If committee members have to negotiate with other committee members and then with members of their own functions or departments, the process will bog down. The net result will be some sort of negotiated compromise rather than a result that really is the best result for the organization.

Third Parties Fill Gaps in Expertise and Transfer Knowledge

Executives' first reaction to this success factor is often to think that it is an endorsement for consultants to come into their organization and run up large invoices. However, that really is not the intent of this success factor. It simply recognizes that in large enterprise projects, such as ERP and also PLM, experience and expertise from having seen and addressed planning issues and implementation problems are critical in an initiative that performs well. In addition, since this is not a one-time event, but a change in the way an organization operates, the transfer of knowledge to the people within the organization is extremely important.

Since PLM is such a new endeavor, chances are that an organization that is undertaking PLM would not have veterans who are experienced in Product Lifecycle Management. Therefore, seeking the expertise of a third party is an example of Lean Thinking. Filling in gaps through third parties will also improve the efficiency of the initiative because mistakes that might be made in developing and implementing a PLM project may have been experienced and solved by these third parties. Also, since the software and other technology that enables PLM is so specialized, the third party resource for PLM might very well be the solution provider who provides the software. However, there are many consulting firms that are developing practices around PLM that claim to have the expertise to help shepherd a successful PLM implementation through an organization.

There are a number of organizations in the aerospace and automotive businesses that engage the solution providers, consultants, and third party process consultants in order to implement their PLM projects. However an organization chooses to do it, it is important that it obtain expertise in PLM because it is a qualitatively different kind of project than they've seen in the past. But at the same time, it is equally important that implementing organizations bring their own people up to speed and have that knowledge transferred to them. Bringing in a third party to implement PLM and then allowing him or her to leave without transferring his or her knowledge regarding PLM is a sure recipe for degenerating back to the old ways of doing things.

Change Management Goes Hand in Hand with Project Management

The importance of project management is not to be underestimated and there is no lack of material regarding project management for the implementers of PLM. However, it is not enough to successfully define, plan, and implement Product Lifecycle Management as it pertains to implementing PLM application software and populating databases with the information and processes from the organization.

PLM is a different way of doing things, especially as it pertains to crossing functional boundaries. Therefore, the change management of the new processes and practices that an organization requires needs to be closely interlinked with the project management. If it simply were a matter of implementing a software package to improve on the efficiency of the current processes within the organization, PLM would not have the potential for the impact that it promises.

The reality is that PLM is a different way not only of viewing the organization's intellectual property in the form of product information, but also of managing that information. Therefore, the processes and practices of the organization need to be reexamined and modified in light of this cross-functional capability. This means that there will be changes in terms of how individuals create and disseminate the information that they are in possession of and changes in how people will use that information rather than re-create it. The organization needs to be looked at from this overall perspective. Rather than simply implementing the processes at hand, the development of new processes and the re-evaluation of the practices that have evolved in order to support those processes need to be assessed and the appropriate changes made.

Again, because the status quo is such a powerful source of inertia, it is important that change management not only be defined, but also closely monitored to make sure that it is not simply a procedure that goes in the book with people going about business as usual, doing it the old way. In order to be able to make this change, the status quo has to be taken off the table, and people need to be prevented from regressing back to the way they have done things in the past.

The Final Success Factor Is That a Satisficing Mindset Prevails

As we saw from the first chapter, *satisficing* is a real word, although not used much outside the world of economists. We have pointed out that when it comes to human activities, especially complex activities, satisficing, not optimizing, better describes what we do. In most human endeavors, diminishing marginal returns are in effect and the effort we spend to attempt to optimize at the end of a project could be more productively spent on the project's next phase. It is not only an issue of resources, which are finite, but in wall clock time, which is invaluable because it can never be recaptured. In the interests of productive use of time and resources, we need to look at getting it "pretty good" and moving on.

ERP and also, by association, PLM fall into the category of very complex undertakings where, if we attempt to try for an optimal solution, we will never get there. Instead, we need to realize that, as we will discuss in the next chapter, we can get to about 80 percent efficiency with human endeavors. Therefore, we need to get the project over a hurdle rate of efficiency, but need not look to optimize it.

There are so many aspects of PLM that if we focus on optimizing one aspect of it, we will let go undone tremendous other opportunities that we simply do not have the attention or resources to attack if we are trying to wring every last little efficiency out of this single aspect of it. With PLM projects, we should set our goals so that they are stretch goals, but ones that are not so difficult to obtain that we will expend all the attention and resources of the organization in a futile attempt to achieve them. Instead, we need to take a larger and a longer view of the situation and look to attack as many areas as possible that can return a sufficient hurdle rate of efficiency for us so that we can improve the overall organization rather than try to optimize only one aspect of it.

Acting Like the CXO

While PLM should be a comprehensive corporate initiative, it is not always the case that the leadership of the organization has been introduced to, recognizes, or has been convinced of the value of

PLM. It may be that the value of PLM is recognized by managers closer to operational issues. The question for these operational managers is "Are PLM initiatives worthwhile if they are not supported at the CXO (that is, CEO, CFO, CIO, etc.) level?"

The answer is that, while the most significant returns are only available when PLM strategies are developed and adopted across the entire organization, even PLM initiatives that are limited in scale and scope can return positive benefits to the organization. In addition, as long as the PLM initiatives support the corporate objectives, which invariably they do, these smaller PLM initiatives can provide proof-of-concept and serve as a model to be emulated by the wider corporation.

As mentioned in Chapter 1, Six Sigma projects can and do bring PLM initiatives to the surface. Six Sigma projects are focused on identifying sources of waste in the organization. These initiatives have the opportunity to be recognized by senior leadership and extended to the broader organization. The multi-million dollar, corporate-wide PLM initiative at Lear Corporation, a $13 billion automotive supplier, started as a Six Sigma project.[6]

There may be some wasted costs if PLM projects are not done as a corporate initiative. Different departments may select and implement different PLM applications. When PLM becomes a corporate initiative, there most certainly will be conversion and consolidation costs in order to standardize the organization on a common technology platform. However, this is far better than the alternative of doing nothing to prepare the organization for PLM. So long as PLM is directionally compatible with the organization's goals and objectives, then undertaking PLM initiatives at the department level or even below is beneficial for the organization.

Here are some suggestions for managers who want to experiment with and gain experience from PLM.

Find PLM Initiatives to Support Corporate Objectives

Since most organizations have as their objectives to grow the organization and to reduce costs, it is not hard to align PLM projects with corporate objectives. From the IT Value Map, we can see that the drivers of revenue growth are increasing functionality, quality, and quantity of product sold. Parts reuse, start parts, and smart parts all allow areas to do more with the same amount of resources. If additional resources are not available to increase the quality or functionality of products, PLM applications enable areas or departments to free up the needed resources by utilizing existing resources more efficiently.

On the cost side, better visibility and control of product information allows the reduction of material and time. PLM applications that track and control math-based designs allow departments to avoid wasting time working with old and outdated versions. Process reuse in manufacturing allows departments to reduce the amount of worker-hours required to produce product routings. Simulation of those routings decreases ramp-up time required to produce products at their most efficient level.

See Beyond Functional Barriers

It is easy to develop a parochial attitude to the area we work with and much more difficult to develop a wider view. It is especially easy in the face of resource constraints to take actions that benefit our own area to the detriment of another area. However, we need to resist such normal responses and take a larger view. An effective PLM strategy will require this larger view and, while there may be short-term benefits in a parochial view, eventually the organization will be less competitive than organizations that do have a PLM view and our own area will eventually suffer.

It is far better to try and develop this larger view of product information sooner rather than latter. We can develop this view even if we do it informally by establishing ties to areas that are adjacent to ours, such as engineering to manufacturing or manufacturing to service. These are areas that develop information that we use or areas that we develop information for. While we may not have the authority to establish cross-functional teams, we can develop informal communications and consultations with these other areas.

For organizations that are highly specialized in a specific function so that electrical, mechanical, and hydraulic engineering are separate and distinct specialties, working on cooperative efforts within the same function area is a reasonable goal. In these situations, PLM can serve as a common focus to coordinate and consolidate product information that all these subspecialties can share and use in one place

Watch for Optimal Decisions That Are Suboptimal

When the pressure is on to reduce resources, the first candidates for elimination are resources that support cross-functional information flow. It is easy for the engineering department to give up resources that support process development needed by manufacturing, or for service and support to eliminate resources to capture and analyze warranty information that could be used by engineering to reduce defects and improve the product.

We can usually try to optimize the use of resources in an area even if that optimization causes a suboptimal use of resources across the entire organization. The problem is that these are easy decisions to make. Making these resource decisions improves our own area and the derivative negative impacts to us from other affected areas are minimal or negligible.

Managers of areas facing this type of decision should resist this opportunity, because these easy decisions go against the spirit and practice of PLM. Instead, they should look for resource usage that is wasteful and only affects their area. In order to provide some visibility for these issues, managers should enlist their counterparts in the organization who will be impacted by the loss of information from potential resource cuts in analyzing the overall organizational effect. By showing that the cost of re-creating this lost information is many times the cost of the resource in the area creating it, there may be other alternatives to eliminating the resource creating it.

At the operational level, there may be "budget trading" between the creating and using departments. With visibility at higher levels, there may be budget relief granted to the creating department. At some level of the organization, there is responsibility for the aggregated budget. Managers need to find that level and demonstrate that PLM information should not be sacrificed at the departmental level at a cost to the entire organization.

Stretch Change Muscle

This is counter-intuitive to most managers' instincts. What all managers think they want to strive for is a smooth running organization where each day mirrors the day before. The people are the same. The operations are the same. The workload is the same. All the manager has to do is sit back and watch it run.

Even if that was possible in the past, the global environment of today means that at some time change will be visited on the organization, whether it likes it or not. It is far better to be proactive with change than let the organization develop a sense of complacency. Organizations that embrace change are more likely to see and adapt to change as it occurs—and, in fact, view change as a competitive advantage. Complacent organizations will be unprepared for change, and if they are able to deal with it all will do so in a reactive fashion.

PLM, with its new approach to products, will allow managers to begin stretching the change muscle of their areas and departments. Getting their people used to—and better still, anxious for—change will allow them to adapt to larger change that comes about when PLM becomes a corporate initiative. In addition, people who have exercised their change muscle will be better prepared to embrace the digital way of doing things and replace the paper they have always used.

Think "One Organization"

Since PLM is an "approach," one of the most important things any employee can do to act like a CXO is to embrace the "One Organization" theme. The issues with organizational dysfunction usually arise when, as discussed above, individual areas think they are benefiting the organization by optimizing their operation. In some cases, it takes on a contrarian theme, "If the rest of the organization is doing X, then we, who are so much smarter, should do Y." It can also occur innocently, and literally without thinking, "We'll do Y because all we have considered is our own little area."

However it occurs, making decisions about how people deal with product information within a certain functional area, what processes and practices are employed, and what technologies are acquired to support these activities without taking into consideration the entire organization is counterproductive. Sometimes, satisficing or getting "good enough" is better for that organization than a misguided attempt at ruthless optimization.

It is not enough for the CXOs to paint a "One Company" vision. Everyone is responsible for using that vision to guide their decisions. CXOs do not have the detailed knowledge about the product information that is used day to day to make all the decisions that an organization needs to successfully capture and use product information throughout the entire lifecycle. Managers and their employees do. Everyone is capable of internalizing the vision of "One Company" as they ask themselves what is the best way for my organization to get the most value out of product information?

PLM has the most impact for the organization if it is a corporate initiative defined by a comprehensive strategic plan that supports the goals and objectives of the entire organization. However, if PLM currently does not have this level of visibility and commitment, the alternative is not to abandon any effort to engage in PLM initiatives that are more limited in scale and scope. PLM can be bubbled up from within the organization. With success comes increased visibility. Finding PLM initiatives or simply projects that support corporate initiatives across functional boundaries, that discourage suboptimal decision making, and that move toward a "One Company" philosophy are all opportunities to show by example the value of PLM.

Summary

In this chapter, we have investigated what it takes to develop a strategy for PLM. It requires a vision of the future, an assessment of where we are today, and a plan for bridging the gap between the two. In order for the plan to be executed, resources and capabilities are required, especially the capability of leadership.

In executing the plan, we can learn from other organization-wide initiatives, such as ERP. There are characteristics that differentiate a successful ERP initiative from a failed one. These characteristics can be applied to PLM. They are: top management is involved and not just engaged; project leaders are veterans and team members are decision makers; third parties fill gaps in expertise and transfer knowledge; change management goes hand in hand with project management; and a satisficing mindset prevails.

Although PLM has its biggest impact as a corporate initiative, there are actions and initiatives that managers who do not control the corporate agenda can take. They can think like a CXO and initiate PLM projects that support corporate initiatives, engage in sharing information with adjacent functional areas, refrain from taking a myopic view of their decisions that affect other areas, and encourage their peers to work toward "One Company."

The impact of PLM is the greatest when there is a strategy that coordinates all the organization's effort. While the returns on this approach may not truly be magnitudes of difference between individual projects, there should be substantial payback for strategy that greatly magnifies its impact over individual projects.

Notes

1. I am, of course, greatly simplifying the elements of strategy. The process of strategic planning consumes a great deal of resources and leadership attention, although maybe too much (see H. Mintzberg, *The Rise and Fall of Strategic Planning: Reconceiving Roles for Planning, Plans, Planners*, New York, Toronto: Free Press, Maxwell Macmillan Canada, 1994). Developing the mission statement, goals, and objectives, performing SWOT (Strength, Weaknesses, Opportunities, and Threats) and VRIO (Valuable, Rareness, Imitability, and Organization) analyses, creating strategies and tactics are all things I learned as an MBA student at Oakland University and practiced as CEO of my own companies. I still pull my MBA textbook from the shelf when I get involved as a board member in strategic planning. However, when thinking about strategy, I use the framework presented here. For those interested in more detail on strategic planning, see: J.B. Barney, *Gaining and Sustaining Competitive Advantage*, New York: Addison-Wesley Publishing Company, 1996 and M. E. Porter, *Competitive Strategy: Techniques for Analyzing Industries and Competitors*, New York: Free Press, 1980.
2. "*Que sera, sera*" was a popular 1956 song performed by Doris Day. The song won an Oscar in the Alfred Hitchcock move "The Man Who Knew Too Much." "*Que sera, sera*" translates to "Whatever will be, will be."
3. See C.F. Gibson, "IT-Enabled Business Change: An Approach to Understanding and Managing Risk. *MIS Quarterly Executive*, 2(2), 104-115, 2003.

4. For the role of adapting by developing an emerging strategy, see H. Mintzberg, "Patterns in Strategy Formation," *Management Science*, *24*(9), 934-948, 1978.

5. See C. Brown and I. Vessey, "Managing the Next Wave of Enterprise Systems: Leveraging Lessons from ERP," *MIS Quarterly Executive*, *2*(1), 65-77, 2003.

6. The Lear Corporation PLM project is detailed in a case study developed by the author for the University of Michigan PLM Overview course. The case study is available at www.corestrategies.com/plm.

Conducting a PLM Readiness Assessment

PLM IS MAJOR SHIFT in perspective for most organizations that have an orientation to traditional functional perspectives. In addition, PLM is not simply an application that is implemented as-is. There are issues of culture, process, practice, and even power that will affect the success of PLM within the organization. This chapter lays out those issues that need to be examined and provides techniques and methodologies to help the practitioner assess his or her organization's readiness for PLM.

Assessing an organization's readiness for product lifecycle management requires an assessment of all the elements of PLM: technology, not only of the enabling PLM technology, but its infrastructure, people, and their processes and practices. This assessment needs to be done using a systematic and understandable framework that compares where we are with where we determine we need to be.

Infrastructure Assessment

As we saw in Chapter 2, PLM is defined as being an approach, not simply technology. However, technology and, more specifically, software are key components of and enablers for PLM. PLM initiatives are not feasible without PLM software applications. These PLM software applications require a computer/communications infrastructure on which to run. In addition, these PLM software applications are not trivial in their infrastructure requirements, so a careful assessment of the computer/communications infrastructure is an important aspect of any readiness assessment.

With respect to computer/communications infrastructure, we noted earlier that computer/communications infrastructure is a hurdle issue. Having excess capacity with respect to computer/communications infrastructure does not add anything to the PLM initiative. However, having less capacity than required can cause the Product Lifecycle Management initiative to flounder and possibly fail. As a general statement, users are comfortable with the computer applications they currently use. They reluctantly use new applications. If the new application lacks minimum responsiveness because of inadequate computer/communications infrastructure, users will employ one of a number of potential coping strategies.

In the least disruptive, users will chronically complain to their superiors, who will begin to question and re-evaluate the usefulness of the application. More disruptive to the PLM initiative, users will circumvent or bypass the system by minimizing their use of the application; by entering only the minimally required data; by batching entry to off-periods, thus delaying the availability of necessary information; or, in the more extreme cases, by abandoning the application entirely and returning to a former or ad hoc method to get their jobs done.

So, in assessing the computer/communication infrastructure, it is important not only to assess what is required under the current initiative, but also to look out over the future and project what will happen to this infrastructure if Product Lifecycle Management becomes a successful initiative (i.e., if people adopt it and change their way of doing things so that they use these new PLM tools and technologies to their fullest extent). With respect to this computer/communications infrastructure, there are four things that we need to look at: the adequateness of the current technology, the scalability of the technology, the modularity of the technology, and the openness of the technology.

Adequateness of the Current Technology

With respect to adequateness, we need to assess the components of the computer/communications infrastructure to see that they are adequate for the current applications that are being deployed on them. If the computer/communications infrastructure is questionable for the current environment, it will definitely be inadequate for the new requirements the PLM software applications will place on it. We also need to assess and evaluate the new requirements that these PLM applications will place on the infrastructure. The three major component areas that need to be assessed are computing capability, bandwidth capability, and storage capability.

With respect to computing capability, the two different aspects are the computing capability with respect to the individual users, and the computing capability with respect to centralized applications and databases. Looking at the individual users first, the assessment needs to take into consideration the current computing requirements of the individual users. If users did not manipulate math-based models in the past, will the new Product Lifecycle Management initiative require them to do so?

For example, engineers who previously dealt only with specifications contained in Excel spreadsheets or Word documents may now bring up a visual model and check its specifications directly, in addition to seeing how various potential modifications affect it. On the factory floor, manufacturing engineers who used to lay out two-dimensional drawings on a table and who puzzled out the three-dimensional geometry of a part may simply bring up the part on a computer screen, flip it, rotate it, deconstruct it, and do other operations on it that would enhance their ability to understand the part and add to its manufacturability.

The requirements of working directly with math-based representations will substantially affect their personal computing requirements. The computer systems that were more than adequate for Excel and Word applications may be completely unsuitable for the compute-intense manipulation of math-based objects. A careful assessment of people affected by the PLM initiative, an inventory of the computer systems they use, and an analysis of the new requirements placed on those computer systems need to be undertaken.

On the server and database side, the requirements for people to access the central repository of models and math-based drawings will put substantial pressure on the resources of those systems to respond. Where requests for information were centralized previously or only used by a small subset of the engineers and designers, the proliferation of math-based representation throughout the organization will cause an increase in the amount of resources that a server will need to service all of these different requests. As a result, the servers may be expected to have an n-fold increase in the service requests that the server receives for searching and delivering the engineering data and math-based representations on demand.

The one thing that is certain regarding Product Lifecycle Management initiatives is that if they are successful, the requirement for computing resources will increase, and most likely increase dramatically. As we will discuss later, whatever computing capabilities are in place will need to be scaleable.

The second aspect of infrastructure is storage. The amount of storage that will be required to store math-based representations in various stages of completion will also increase dramatically. In addition, depending on the policies with respect to the organization on backup and security, it may be a requirement not only to have the storage occur in one place, but also duplicated in multiple places.

This is also true if it is impractical to store drawings in one place because of the bandwidth requirements to serve them to diverse geographical locations. The requirement may be that there are servers in different geographical locations that will serve up these math-based drawings to the users. In this case, these servers need to be kept synchronized and, as a result, the storage requirement increases will be fairly dramatic.

Finally, the bandwidth issue needs to be assessed carefully inside individual areas of the organization, in different geographical locations within the organization, and outside of the organization. Moving math-based drawings currently takes a tremendous amount of bandwidth. If we give access to a wide variety of people who currently do not have this ability, we would expect to see the requirements for bandwidth to increase substantially.

In addition, transfers of math-based information between geographically dispersed locations will also increase. If the interest is in part reuse, the ability to move those parts from other parts of the organization and gain the advantage of the work that has already been done on developing these parts will cause bandwidth capacity to be used for this purpose.

Scalability of the Technology

Special care must be taken to look at the upper limits of computing, storage, and bandwidth, and determine whether there is any point at which the system will become fully saturated and unable to scale. That point must be assessed fairly carefully because, if it is not, the infrastructure will build to that particular point and then degrade rapidly, putting the entire system in jeopardy.

Modularity of the Technology

Modularity is important because it makes sense to add capacity in small increments. In the days of the mainframe, the step function for adding capacity and its associated cost was huge, and sometimes even prohibitive. The only recourse when a mainframe computer system became saturated was to buy a new mainframe and split the load. With the cost of mainframes in the millions of dollars range, it was often the policy to let the system degrade as far as possible before adding another mainframe.

With the computing power of today, the cost to add a new increment of computing capability or storage capability is generally not that great and, therefore, that decision is made fairly easily. Although some scarce computing resources, such as special purpose computers to do heavy duty simulation, can be expensive resources, their use is limited to a small area of experts and can be allocated fairly efficiently to get the maximum utilization.

The network backbone is generally the constraining factor for bandwidth and increasing its capacity is usually an expensive proposition. There are some techniques that can be used to change the modularity and improve the capacity of the backbone infrastructure. One technique is separating out users who share files on a constant basis and localizing their file sharing. This is so that the files are not required to go over the entire network. Another is to send only changed data, such that the information that has changed needs to be updated and not the rest of the file, which may be unchanged.

Openness of the Technology

Finally, openness will be extremely important for Product Lifecycle Management, since the scope of PLM is so vast. No one solution provider will be able to perform all of the functions that are required in an organization for full Product Lifecycle Management implementation. Open architectures on both the hardware and software side are a must so that technologies and applications can work together to provide the functionality that the customers require.

The first best solution is obviously to have interoperability with common data formats and common data models to facilitate the exchange of information. The next best solution is to provide conversions or translators to convert and translate the information into a common format that can be used by different programs. XML (Extensible Markup Language) and STEP (Standard for the Exchange of Product model data) specifications are examples of common formats that solution providers may support that would provide openness to the user community. It is doubtful that a PLM standards effort would succeed or even be desirable. However, openness does allow for harmonization where different PLM applications can coexist and share information.

At the 2003 University of Michigan AUTO^e Conference in Detroit, Dan McNichols, Vice President of Information Services and Systems for General Motors, outlined General Motor's approach to preparing the infrastructure for a digital environment that could support PLM. The backbone of his infrastructure was a high-speed data network that could connect all his facilities, no matter where they were in the world. GM also replaced older computers with high performance computers so that the engineers and other users of PLM information were not constrained by computing capacity. His requirement was that the system grow openly, scalably, and modularly so that GM could add additional capacity as required.

Assessment of Current Systems and Applications

It is important to assess the current systems and applications that are in place within an organization. The easy analysis is to assess the formal systems that are in use. Information technology groups or IS staffs have this information available, since they are generally responsible for installing, maintaining, and supporting these types of systems. Generally, the systems are well documented, and a map of the information flows within the system and from system to system is readily available for analysis.

The issue becomes assessing the informal systems that have developed within organizations to provide information on particular tasks or subtasks. This information is commonly in Word or Excel spreadsheets and exists on local servers or is moved from user to user via e-mail attachments. There is nothing wrong, per se, with these informal systems other than that they are not very visible to the organization as a whole. In fact, these informal systems may contain critical information regarding products and may facilitate and expedite critical decisions regarding product structure, cost, or other needed information.

In doing an assessment, it is important to perform an inventory of these informal systems and assess them to determine the compatibility of the new system with these informal systems. At a minimum, these Word documents and Excel spreadsheets need to be collected and organized by the new PLM system such that the information is all in one place, even if the formats are incompatible or unknown.

It is also extremely important to look at the development of these informal systems between functional areas. These informal systems generally fly under the radar of the information systems staff, and as a result may not be identified and captured during the assessment process.

Because systems have tended to be siloed, the movement of information between functional areas can tend to be ad hoc and may have developed as a practice over the years. For example, the purchasing department may be getting additional information about the product specifications from the engineers as they make their decision with respect to vendors. The manufacturing group may get alternate configurations from the engineering group so that it can assess the best way to manufacture the product and feed that information back into the engineering group prior to the final design and its associated Bill of Processes being approved.

In some cases, this information is not conveyed in informal systems, but may be conveyed in the formal system itself. However, the information may be put into comment fields or fields that are thought to be free-form but really have specific formats that have been developed through the practice of these different functional areas. The IS staff may be unaware that structured information flow is taking place between functional areas.

The issue as to whether the new PLM systems that are being installed will be replacement systems or will be integrated within the current in-use systems is very important. If the system is a replacement system, it is much more critical to understand all the information flows than if it is going to integrate into current systems. These informal systems will wreak havoc with the replacement system because the accepted practices of information transfer are suddenly replaced by this new system. The way people have done their jobs in the past is disrupted because they no longer have the same information flow.

If the new PLM system is going to integrate with current systems, then the problem is lessened because, as these informal systems are identified, they can be linked or integrated into the new system. So, for example, if the system will integrate word-processing documents and Excel spreadsheets, then the appearance of a few more of these that people are not aware of, at least on an official level, is relatively simple to handle. However, even in these cases, it is important to identify these new and unknown informal systems because it gives the PLM implementation team the ability to fully map the information flows within an organization.

In any case, the purpose of this assessment is not simply to assess the formal systems that are in place within the organization and are all very nicely defined, but to get to the underlying and informal systems that exist within the organization. There are two reasons for this. The first is to make sure that any new system will incorporate these information flows within it. The second is to get the opportunity to assess these information flows and see if the practices that has been built up over the years ought to be defined as a processes.

The problem with these practice developed systems is not that they are inefficient, because generally they are efficient. These practice developed systems are efficient because they have been developed by people who are trying to do their jobs better and have probably given a lot of thought to the situation. The problem is that they are undocumented and not formally known to the organization, so that if the people who are capturing, maintaining, and acting on this information are replaced or retired, that information is suddenly lost. The well-functioning operations that have existed up to that point in time are suddenly disrupted, and no one may know why.

An example of this was the paint department of a major manufacturing firm. A long-time employee used to monitor the makeup of the paint and the weather conditions at the plant. When the temperature was predicted to dip below a certain point, this employee would order a preservative chemical mixed into the paint to prevent the batches of paint from freezing. When this employee retired, the batches of paint that were produced in the winter suddenly froze with no one in management knowing why. It was not until the management investigated and talked to the employees who had worked with this retired employee that they found out that he was performing this additional process without anybody taking particular notice of it.

While this process was effective as long as that employee held his position, this information from this informal system was suddenly lost upon retirement. Because the retiree's replacement did not have the same informal relationships with either the information or the coworkers, he or she was unable to make the addition of preservatives to the paint, to the substantial economic detriment of the corporation.

People Assessment

With respect to product information, one of the first assessment opportunities is to determine the amount of time that has been spent on unproductive tasks. What an organization needs to get its hands around is how much time its people spend on searching, duplicating, re-creating, copying, distributing, and simply wasting information.

The estimates that are commonly quoted are that somewhere between 60 to 80 percent of an engineer's time is spent on unproductive tasks as listed above. The amount of time that engineers spend on these unproductive tasks can be assessed through surveys, interviews, and observations. Good methodologies exist to determine the percentage of time that engineers, designers, and others spend on these unproductive tasks.

With respect to the tasks of duplicating, re-creating, copying, distributing, and wasting information, the waste of those tasks are relatively straightforward and are clearly unproductive uses of people's time. With respect to task of searching, however, we need to be careful to assess whether searching is productive or unproductive. If the issue is searching for information that is specific and known to exist, but difficult to find, then we can all agree that this type of searching is an unproductive task. However, in many cases, when people are searching for information, they are also assessing and creating within that process.

So, while an engineer may look as if he or she is searching for information with respect to a specific part, what the engineer also may be doing is assessing that part's suitability for the functions that he or she has in mind. In some cases, there are very complicated trade-offs that require the engineer to look at a number of similar parts to determine what specific trade-offs he or she really would like to make and how those trade-offs should be made.

This is not an unproductive search, but, in point of fact, is a highly productive search, in which the engineer is using his or her intelligence about various aspects of the product design and the context in which it is being designed to make subjective but critical judgments. This is not a task that will be replaced simply by giving the engineer better information in a Product Lifecycle Management system. In point of fact, the Product Lifecycle Management system will need to have rich enough search capabilities to provide the engineer with the information that he or she really needs. Now it may be that the actual search aspect of it can be done more efficiently. For example, the engineer can do the search at a computer rather than having to look through drawings in a drawing room. However, a great deal of the search process is searching for something that engineers don't know they are looking for, as opposed to a specifically defined piece of information.

Once an estimate is made on the percentage of inefficient time individuals spend on searching, duplicating, re-creating, copying, distributing, and simply wasting information, then an assessment as to the cost benefit of a Product Lifecycle Management system can be developed by multiplying the workforce engaged in those activities by that percentage. This impacts the cost aspect of the IT Value Map and represents quantifiable cost savings.

The second aspect that needs to be assessed is the magnitude, ability, willingness, timing, and distraction of the people involved in adopting a new system.

With respect to magnitude, people can handle smaller magnitude changes better than they can handle large magnitude changes. In addition, the magnitude of change needs to take into consideration that there is a transition period in learning a new system, which amplifies that magnitude during the period of time that the learning takes place. To the extent that this magnitude can be lessened by phasing in the system, adding additional people to help during the transition period, or other techniques, magnitude can be lessened for the people involved.

The next assessment concerns the ability of the people to handle the new system. On the information technology side of the house, the assumption is often made that all people are as comfortable with new systems as information technology people are. The assumption is also made that all people have the ability to handle different systems as easily as information systems technology people do.

However, there is a large group of people in any organization who view new systems with fear and trepidation, and this substantially impacts their ability to deal with new systems. The assessment that needs to be made is the ability of the people to handle these new systems, and that assessment requires understanding the amount of training and education that needs to be given. This also needs to take into consideration the fact that people generally continue to have a job to do as the new system is rolled out, so the luxury of taking people off-site for extended periods of training time is generally not available to most organizations. An assessment of the trade-off between longer implementation times and more focused training needs to be made along with this assessment.

The willingness of people to make changes can be a major stumbling block. If people are very comfortable with the status quo, or are rewarded for the information that they know that is unknown by their coworkers, this will substantially decrease their willingness to move to a Product Lifecycle Management system, where information is shared freely among the workers that need it. In addition, if the organization is structurally set such that there is competition between functional areas and that there are winners and losers within the organization with respect to possession of information, this will also negatively impact the willingness of people to share their information.

This willingness is involved not only with the actual implementation of the system, but also may be a factor in the assessment. People may hold back information, mislead about the information that they need for their work, or not admit to informal practices if there is a strong culture in place that emphasizes only the formal systems. In addition, where there are power struggles with respect to control such as in a strong management versus a strong union shop, the issue of people's willingness needs to be examined very carefully.

With respect to assessment, we also need to look at the issue of timing. We may assess that people have the magnitude, ability, and willingness to adopt this particular system, but we are unaware of the fact that they are being asked to adopt three or four other systems in the same period of time. If we are piling change upon change on the people who are being asked to adopt these systems, we may find out that they simply are unable to absorb all the changes at one point in time. In fact, we can probably make the determination that this will be the case. As a result, we need to make the assessment not in a vacuum, but with full awareness of the entire portfolio of different systems that are being proposed and adopted at the same point in time. We must observe this and be able to space out the timing of the changes in systems that we're asking people to absorb.

Finally, we need to assess the elements of distraction. Distractions are those events within the organization that may detract from the attention level of the people who are responsible for these new systems. If there are layoffs that are rumored to occur, or reorganization of departments, or even new management being brought in, these are all things that distract people from the day-to-day operation of their business, to say nothing of the implementation of a new system. While there is some level of distraction that generally occurs in any organization, such as people moving into or out of the organization and new business opportunities being sought after or lost, if the level of distraction is unusually high, we need to make an assessment and understand that assessment prior to trying to implement a new system. We obviously should try to minimize any distraction or look for a period where the distractions are going to be at a lesser point before asking people to focus on change and adopt a system that alters the way that they have done their jobs in the past.

Process/Practice Assessment

An assessment of processes and practices needs to be undertaken in order to understand exactly what information is needed and used within an organization. With respect to process assessment, the requirement is to verify the completeness of process maps, evaluate gateway placement, and map paper to digital processes.

Most organizations develop very complete and detailed process maps. However, they are generally obsolete the day that they are developed because process maps are static while the organization is dynamic. Everyday processes are adapted and modified to fit the needs of the new requirements that have developed. As a result, if process maps have not been reviewed for a while, they can be hopelessly out of date. Any system implementation that is based on the state of these process maps is bound to become mired down in issues before it is completed. While it is a good idea to update process maps on a regular basis, it is especially imperative to make sure these process maps are up to date when implementing a new system.

The next aspect of process assessment is to evaluate the gateway placement. Decisions about evaluating and approving product milestones require that there are gateways for this product to get through in order to maintain the viability of the product process. However, sometimes these gateways are formally defined but, in practice, are replaced by more informal methods. While the gateway process may call for a formal review by a particular committee at a particular point in time, people learn very quickly not to bring products that will not be approved by that committee. What develops are informal processes to assess that product at earlier stages in its development so that only products that have a certain percentage of success will be brought through that gateway process. In other cases, the gateway approvals are after-the-fact formalities. An assessment of gateway placement needs to ensure that these gating events are real, operational events and not formalities, either before or after the fact.

Finally, assessments need to be made about mapping paper to digital processes. In many organizations, paper is a very comfortable thing to move around from person to person, and simply holding paper as one would hold a token becomes part of the imbedded processes within an organization. Removing that piece of paper changes the dynamics of how people work. As a result, before that paper is replaced, the digital process that is going to replace it needs to be identified and understood. In addition, we need to make sure that that paper does not contain other information, such as notations or other comments that people make on it in the normal course of performing their tasks, that will not be incorporated into the digital system. If that is the case, we will find that people will use the paper system along with the digital system because the digital system does not contain all the necessary information that they require.

While processes are fairly straightforward, practices generally are not. The three elements of assessment are to develop interface maps, to map information streams and wells locations, and to identify rote practices. Developing interface maps means understanding who interfaces with whom in the organization. It is only by looking at these patterns of interaction that we can determine whether or not informal systems have sprung up that will not be reflected in the formal system that we are attempting to replace. We may think that the interface for a particular department only exists at certain points and with certain designated individuals. In addition, we may believe that this department only communicates with certain other departments. However, if we examine the communications patterns we may find that there are different departments that interface or different individuals that interface other than the previously identified ones. We need to identify all these interface points to fully understand how people go about processing information for their jobs.

We also need to map information streams and wells. By streams we are talking about the flow of information within the organization on both a formal and informal basis. Wells are those repositories of information that exist to provide information that may be more difficult to obtain or is not readily available. A well of information can be a database, either internal or external, that people refer to in certain situations, or it may be a person who has a deep repository of knowledge in a certain area whom people go to when the normal information streams are insufficient for their particular jobs. These information wells need to be identified and incorporated into any plan that we have for information with respect to the product information that we are dealing with.

We also need to identify rote practices. These are practices that began as practices because people were unsure of how to link inputs to outputs. They are now understood well enough to be formalized in a process. However, since they have always been done as a practice, no one thinks to formalize them.

An example of a rote process would be to have an inventory control person visit each station daily to assess inventory on hand, review the daily production schedule, determine the amount of inventory required at the station, and place an inventory requisition. Since this information is available in the production system, this could be made into a process and done automatically.

This process and practice assessment needs to take place in order to be able to ascertain whether the product information that we are thinking of providing will be sufficient for the people to do their jobs. If it is not, we will find out very quickly because people will redevelop these processes and practices by bypassing the system and we will lose that information for other uses and benefits.

Capability Maturity Model Assessment

Once we have identified the areas we need to assess, it is important that we establish an assessment framework. One that appears to have gained acceptance in the information technology area and has been used by solutions providers in their assessments is the Capability Maturity Model (CMM). The Capability Maturity Model was developed at Carnegie Mellon University's Software Engineering Institute (SEI). CMM was originally designed to assess the maturity level of software development organizations. However, the framework has since been adopted for other uses, such as PLM assessment.

The Capability Maturity Model has five levels. They are:

1. Initial (ad hoc)
2. Repeatable
3. Defined
4. Managed
5. Optimized

At the initial level, all the work an organization does is done on an ad hoc basis. Like our quasi-amnesiac tribe, for these organizations every day is a brand-new day. There are very few processes in place, or at least the processes that are in place are not repeatable or definable. Generally, companies that are at the ad hoc stage are very inefficient and rely solely on the capability of their people to address the tasks required. The better the people they have, the better these organizations get by. The worse the people they have, the more difficult it is to get by.

The second CMM level is repeatable processes and practices. At this stage, the organization tends at least to do the same things in the same way, given the same specific inputs and the desired outputs. People have discerned a pattern and are basically repeating that pattern on a day-to-day basis. It is output-efficiency based, meaning that it appears to work, so therefore people tend to do the same things over and over again. Every day is not a brand-new day, and people have learned and retained that learning about the tasks they have to accomplish.

The third CMM level consists of defining these processes and practices. We now have enough information about them to define what the inputs are, what the desired results or outputs are, and what processes and practices are needed to link inputs and outputs. We can document this so that we can compare this situation to future situations to ensure that it is repeatable, But, more importantly, we can institutionalize these practices and processes so that new people coming into the organization can understand what needs to be done in a much shorter period of time.

The fourth CMM level is to manage these processes and practices. We analyze the processes and practices and develop measuring methodologies to determine the amount of resources the processes consume and the results or output we obtain. We develop specific goals and objectives for the resource consumption and output quantity and quality. We then measure each time we do the processes and practices to see how we do against our goals and objectives.

The fifth CMM level is to optimize the processes and practices. We not only measure and evaluate the processes and practices against our goals and objectives, but we look at ways to continually improve the processes and practices so that our goals and objectives increase. We continually look at these processes and practices and not only assess them, but also their environment, so we can continually adapt them to the changes in the environment. It is at this optimized level that we are also being proactive in looking at potential changes coming down the line in order to be able to assess the systems, processes, and practices we have in place and make sure that we are changing the organization prior to their occurrence rather than reacting to them after they occur.

The levels of Capability Maturity Model are anchored to these specific attributes and should not be confused with an arbitrary ranking of 1 to 5, which is commonly referred to as a Likert Scale. As anyone who has taken surveys knows, the Likert Scale is an arbitrary scale for ranking preferences in which the coding usually is 1 for "strongly disagree" to 5 for "strongly agree," with 2, 3, and 4 in the middle corresponding to "moderately disagree," "neutral," and "moderately agree." Survey takers indicate their preference on this 1 to 5 scale, and it is both arbitrary and relative to the respondent in nature. The Capability Maturity Model, on the other hand, is not arbitrary and should correlate to specific attributes as outlined above.

If we look at all organizations, they will fall roughly within a range of 30 to 80 percent in terms of efficiency.[1] While it's certainly possible for an organization to be below 30 percent in efficiency, the reality is that for-profit organizations below that degree of efficiency probably do not stay in existence for very long. At the other end of the scale, it is difficult to have a human based organization that has better efficiency than the 80 percent level. However, between the 30 percent and the 80 percent levels is a fairly wide range, and we would expect that there are organizations that fall within all areas of that range.

If we look to plot the Capability Maturity Model ratings against the organizational efficiencies percentages, we would put organizations that are at level 1 (initial or ad hoc organizations) close to or at the 30 percent mark. The organizations that are ranked 5 (organizations that are at the optimized level) would have efficiencies at or about the 80 percent range. If we space out levels 2 (repeatable), 3(defined), and 4 (managed), we come up with a graph that looks like Figure 10.1.

It is important to note that the graph in Figure 10.1 is not linear. The movement from CMM level one, ad hoc, to CMM level 3 is probably relatively linear. The improvements are gradual and evolutionary as organizations make routine and institutionalize what they do. However, there usually is not a great deal of analysis given to the efficiency of the routines. The repeatable routines are ad hoc routines that seem to work well. These routines are then defined because they have been repeated so often.

Efficiency should accelerate as we move to the managed level. We begin to bring to bear analysis and metrics to our routines to

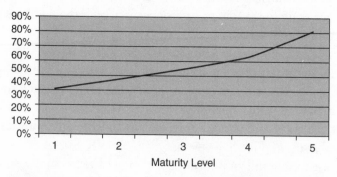

Figure 10.1 Capability Maturity Model Efficiency Effect

actively improve the efficiency of our processes and practices. The largest increase in efficiency comes as we move from the managed level to the optimized CMM level. Our focus here is in continually improving the processes and practices of our organization and continually setting our goals for efficiency higher.

While this graph is only a hypothesis of what organizational efficiency looks like and is not based on any empirical data, it is plausible and certainly illustrates the idea that there are major improvements to efficiency available if companies improve their organizational maturity. The more organizations make themselves more capably mature, the more efficiency they can obtain.

However, it is important to note that these CMM levels are not simply ordinal and arbitrary ratings that are given to organizations. These ratings correspond to behavioral characteristics of organizations and they need to tie to specific demonstrations of capability in order to properly rate organizations on the Capability Maturity Model.

However, all organizations exist as a collection of discrete functions, which we often call departments or areas. To properly assess an organization we need to look at the individual areas and not simply the organization as a whole. So, in assessing our organizations, we must look at the individual functions, departments, and areas and assess their CMM levels individually. As noted below in our discussion of assessment cautions, if the CMM level we assess to the functional area is the average of wide variability within individual areas, it may not be particularly useful.

The performance indicators should be tied to characteristics that allow us to distinguish the various stages of the Capability Maturity Model on the basis of the type of performance characteristics and metrics that we're looking at. However, since Product Lifecycle Management is a corporate issue, we need look at the mapping of the organization in an overall perspective, maintaining the individual functional areas so that we can properly assess them. We can use the PLM Model that we introduced in Chapter 2 to help illustrate the assessment on an organization-wide basis.

If we look at Figure 10.2 we have the PLM Model that we developed in Chapter 2. However, we have replaced the informational core with an axis or vector of each of the functional areas.

Along that vector we have five segments identified for each of the maturity levels. If we designate the center of these vectors to indicate a maturity level of 1, ad hoc or initial processes, and the segment that is closest to the functional area to indicate level 5, optimized practices, we can plot the Capability Maturity level of each of our functions along each vector.

If we connect the dots and shade in the area within the dots, we come up with an area map as seen in Figure 10.3. Based on this kind of visual coding scheme, the more area that we have shaded, the higher the maturity level of the overall organization will be. An organization with little area shaded is obviously an organization at level 1, with little more than ad hoc processes. An organization that covers the maximum amount of area within the figure would have its functions consistently at level 5 and be a highly optimized organization.

We should resist the easy conclusion that the goal of our organization is to have the maximum amount of area covered, indicating that our entire organization is at level 5 and highly optimized. The reason for this is that we do not know without careful analysis and evaluation whether or not all functional areas need to be at level 5. It will all depend on the nature of the tasks and the importance of these functions to the overall organization.

For example, if we have only a single customer who gives us all of our orders for the year and who defines exactly what it is that we

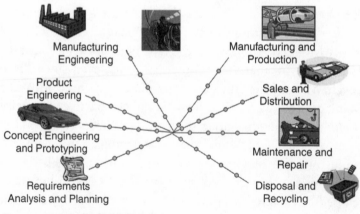

Figure 10.2 CMM Model

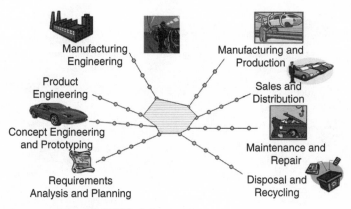

Figure 10.3   PLM Model

need to do to fulfill those orders, then bringing our sales and quotation function to a level 5 would be a waste of time and resources. In this situation, it does not matter that we have ad hoc or initial processes in the sales and quotation area because the customer defines what we are going to produce. We simply have to agree to the customer's request. Therefore, putting resources and effort into bringing our sales in quotation area to level 5 would be a waste of resources that we could spend on other efforts.

Or, to take another example, our manufacturing process consists of a simple assembly operation for an unchanging product. The assembly operation is so simple that there is only one way to perform it. A brand-new employee becomes proficient in a matter of minutes. Going through the process of taking this function from Initial to Optimized is a waste of resources.

While these are extreme examples, the point is that we need to assess each of the functional areas and determine what level it is with respect to Capability Maturity Model and decide on what level we wish each functional area to achieve. If we plot both points and shade in the area as we do in Figure 10.4, we now know the scope of our task. Our task is to close the gap between where we are today and what we wish to achieve tomorrow.

In determining which function to attack first, we should look for the greatest economic impact. While moving one function from level 4 to level 5 might have the most percentage efficiency increase, moving another function from level 1 to level 2 might have the most

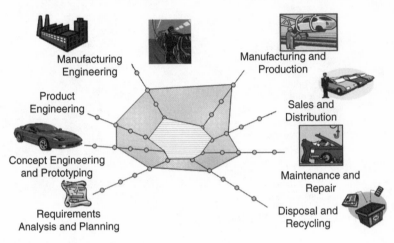

Manufacturing
Engineering

Product
Engineering

Concept Engineering
and Prototyping

Requirements
Analysis and Planning

Manufacturing and
Production

Sales and
Distribution

Maintenance and
Repair

Disposal and
Recycling

Figure 10.4   PLM Model

economic impact if the level 1 function is substantially larger in size and resources deployed.

The second step is to develop assessments for each of the separate practice areas that make up each functional area. The quantitative measures should be validated against efficiency measurements. As we have noted repeatedly, if we cannot show that we can impact the IT Value Map by moving CMM levels, then the initiatives to do so will have difficulty in being approved. Qualitative measures should also be meaningful and differentiate the different CMM levels. Figure 10.5 has a sample practice assessment that has the five levels and both qualitative and quantitative evaluation criteria measured against those levels.

This is a Capability Maturity Model for the product-engineering department in the area of component reuse. This has been developed for illustration purposes only, so it has not been validated against actual practice, although the criteria should be directionally correct. In addition, depending on the type of product development, the quantitative measurements could vary substantially. While organizations can compare themselves to other organizations with comparable product development requirements, there is no generic "best practices" that can be applied universally, although there are certain to be Healthy Practices. However, the qualitative characteristics will be consistent since they are indicators of the processes and practices that are being employed.

Maturity Level	1 Ad hoc	2 Repeatable	3 Defined	4 Managed	5 Optimized
Qualitative Characteristics	Engineers given specs and budget No standards for reuse Easier for engineers to design new than reuse	Repository of designs created Engineers trained in searching repository Reuse is dependent on engineer and accidental	Policy and process in place to search repository for designs BOM reuse scoring performed Reuse is suggestion	Monitoring of reuse performed Statistics provided monthly to management Where-used reporting Engineer bonused on reuse	Automated searching for designs matching specs and budget Automated searching of BOM for similar components Reuse improvement group increases goals
Quantitative Characteristics					
BOM ReUse Percentage	<1%	3%	5%	12%	20%
Engineering Cost/ Revenue	55%	52%	48%	42%	33%

Figure 10.5 Reuse Maturity Assessment

The initial CMM level is characterized by ad hoc practices. Engineers are given product specifications and a target budget and are turned loose. How they develop the design is up to them. Since even the most primitive tribes have common practices, there are typical ways that these engineers will go about their development. However, there will be wide variations depending on educational background, the practices of engineers they work directly with, and the specifications they have to work with. Any component reuse is accidental. It is easier for the engineers to design de novo than to look for components to reuse, unless the designs are a variation of a current product. Inefficiency is high. The percentage of design costs to revenue is substantial.[2]

At CMM level 2, the repeatable level, the organization has developed some repeatable processes. It has created a repository of component designs. Engineers routinely, although not always, deposit their designs in the repository. There is at least someplace to go to look for reusable designs if an engineer is so inclined. Engineers are also at least trained on how to search for component reuse as a practice that is specific to the organization. However, component reuse is still very dependent on the local practices that have been developed by engineers working on those designs. Reuse is still fairly accidental. There is some component reuse simply by virtue of having the repository. The ratio of engineering costs to revenue is still high.

At CMM level 3, the defined level, the organization has formulized the search for reusable components in its engineering process. A repository exists. Engineers have been trained to use it. A policy manual defining how to search for components to reuse has been written and distributed. The organization may have developed a scoring system to define what percentage of the new product's Bill of Materials has been reused. However, it is still up to the engineers' practice to determine whether they will try to reuse components. Reuse is more of a suggestion than a practice. Component reuse increases partially as a result of a reuse percentage that has some visibility. Because of that, the costs of product design to revenue will decrease.

At CMM level 4, the managed level, the organization gets serious about component reuse. Management monitors the statistics for reuse and gives bonuses to engineers for this behavior. The where-used reports are looked at to see not only where the components have been reused, but where they could be reused. The component reuse statistic reflects this measurement and control activity by exhibiting a sharp jump upward. Engineers spend less time on designing components that already exist, so design costs to revenue improves.

At CMM level 5, the optimized level, the organization works constantly on improving product reuse. Automated assistance is provided for searching out component reuse activity. In addition, the product undergoes a formal review to determine which components can be combined. Can two different kinds of bolts be replaced by one throughout the product? The organization has working teams to look at design practices and promote component reuse. The organization is not content with static goals for component reuse, but continues to raise the bar.

With this laser-like focus on component reuse and its continual improvement, the percentage of component reuse rises dramatically and should continue to rise. The engineers' efficiency and the ratio of costs to revenue will also dramatically improve, although the organization may choose to take some of those efficiency increases and apply them to innovating new, different, and more profitable products.

This is not a perfect example of a Capability Maturity Model assessment in that it is unclear how the numerical evaluation numbers correlate to efficiency levels. However, the quantitative measures seem directionally correct and the qualitative aspect seems to tie into this idea of Maturity Level Models. This is evidenced by the fact that in the first stage there is no system in place. In the second stage, at least there is some sort of repeatable process. At the third level, policies are being developed and disseminated. Goals are being managed to at the fourth level. At the fifth level, automated tools are being brought to bear and continual improvement is the focus.

Again, this is just an assessment of what CMM level we are at. There is no value judgment about the organization based on this assessment. An organization at CMM level 1 for a particular function is not necessarily worse off than an organization at CMM level 5 for that function. Thus, we need to be careful when setting up our assessments that we don't try to prejudge the value of the various levels to organizations and only describe how they match up against the various CMM levels.

In our example of the low volume sales and quotation area, even though it may be operating at a 30 percent efficiency level, it may spend so few resources that the difference between 30 percent and 80 percent in terms of recapturing those resources for the organization might be meaningless. Perhaps it might better spend its time to develop better systems in other areas where it has higher volume and the opportunity to impact the efficiency of the organization with a much greater economic impact.

However, this sample practice assessment is beneficial from the standpoint that it shows both a qualitative and a quantitative approach to the assessment issues and it also lays out criteria for making those assessments. As we noted before, we need to be careful that when we develop assessment standards we link them to CMM level criteria and not simply make them subjective and relative one-to-five scales. To the extent that the criteria are numerical, we can use them to examine the economic impact of moving from one maturity level to another. For example, if our product engineering costs are $100 million, ad hoc product reuse is zero, and repeatable product reuse is 3 percent, then moving from ad hoc to repeatable has an economic benefit of $3 million.

The process of developing the appropriate maturity level criteria, assessing the organization against these criteria, and determining the maturity level the organization needs to acquire to meet its goals and objectives is not a trivial undertaking.[3] The assessment can take place at various levels of detail. However, the higher the level at which the assessment is performed, the less actionable the assessment is. This means we may be able to assess our engineering department as ad hoc versus repeatable. However, to be able to move engineering to the repeatable level, we need to assess each practice and process in the engineering department to define at the team level the criteria that we can measure its progress against.

Assessment Cautions

We must point out a number of assessment cautions in order to make sure that we do not make incorrect conclusions from the assessments that we perform. The assessment cautions involve relying on simple math, watching for large deviation and outliers around the practice average, looking for chained inefficiencies, and the effect of externalities.

The caution regarding simple math is that it never works out that the increase in efficiency multiplied by the number of people in the department is the projected reduction in headcount. If the increase in efficiency is expected to be 10 percent and there are 500 people in the department, the headcount reduction will not be 50, at least initially. In the best of situations, improving efficiency takes time and, in the initial stages, there actually may be a decrease in efficiency because people are either using parallel systems or simply adapting their way of working to the new system. Too often, organizations make large promises regarding PLM on the basis of simple math, only to find out that the promised savings do not immediately materialize. As a result, the project gets a black eye.

The second caution is to watch for large deviations and outliers around the practice average. As noted above, the problem is that there may be a diverse group of activities that occur in a large practice area. As a result, we might get a wide range of assessments with respect to the different practice areas. If we average them out, we will mislead ourselves as to what our maturity level actually is. We might have some practice areas that are at level 1 on the maturity model and an equal number at level 5. On this basis we conclude that our average is 2.5, which is completely inaccurate. We have level 1 areas and we have level 5 areas, but we don't have any level 2.5 areas. We have to look at the different assessments for large variability among the different areas. We need to make sure we do not mislead ourselves by simply averaging the assessments out.

Outliers are a similar problem. We assess that most of our practice areas are at level 3, but we have one critical area at level 1. We will mislead ourselves if we conclude, on the basis of averaging the numbers, that we are slightly under level 3. We need to look at the outliers in each particular area and their magnitude of impact on the overall organization. If necessary, we may need to weight the assessments if some are much more important or critical to the organization than others.

The third caution is to look for chained inefficiencies. This is where we fix one of the areas, but because its output feeds another area with its own inefficiencies, we do not pick up any increased efficiencies by improving the first area. An example of this is when we implement component reuse to speed up the design cycle, but the design needs approval from the manufacturing department, which only reviews designs once a month. When doing assessments, we need to look at the weakest link in the chain and attack that area first. Otherwise, we will expend resources with the expectation that they will show efficiencies that will never materialize because they are negated farther downstream.

The fourth caution relates to externalities. These are areas that are outside of our control. Externalities can easily negate efficiency improvements. As an example, we improve our design process and require that we turn around design changes in five days, but one of the steps requires us to send them to the customer for approval. However, the customer sits on the change for 10 to15 days. We can think that we're at level 5 all we want. However, as long as the customer is a bottleneck for this particular process, we will never be able to improve our efficiency. We need to look at what externalities affect the organization and how that impacts the processes and practices we are assessing.

Summary

In this chapter, we have looked at the issues involved with assessing where the organization stands today with respect to its processes and practices regarding PLM. We need to first look at the infrastructure, since that is a hurdle issue. The technology required for PLM is computer-, storage-, and bandwidth-intensive. An organization needs sufficient infrastructure for its current PLM initiatives and, if PLM is successful, those requirements will increase. Scalability, modularity, and openness of the infrastructure are important.

An assessment of current systems and applications, especially informal ones, needs to be carefully analyzed. People need to be assessed for their willingness and ability to change. Processes and, especially, practices require attention. Information flows and wells need to be identified and mapped.

The assessment needs to be done in a framework. We have proposed a model based on the Capability Maturity Model, with its five levels tied to specific characteristics. The organization has to determine where it needs to be and where it is, and then address that gap. Finally, there are assessment cautions concerning simple math, large deviations and outliers, chained inefficiencies, and externalities.

Notes

1. Admittedly there is much fuzziness surrounding this contention. Efficiency is defined as the actual divided by the potential. If we look at the average organization, people work about 2,000 hours per year. Eighty percent efficiency means that they only are productive 1,600 hours per year. Add up lunches, coffee breaks, leaving early for the kids' games, doctor appointments, and 80 percent efficiency looks reasonable. How much people accomplish in the 1,600 hours is another question entirely. If we look at efficiency not as hours but as actual work accomplished divided by potential work accomplished, we get the variance among organizations, even though there might not be much variance in hours worked. How organizations are organized and how mature their processes and practices are will dictate how much actual work they can accomplish. At the high end, human productivity would seem to be roughly in the 80 percent range and still be called human activity.
2. This assumes that not all competitors are at level one. Where there are more efficient competitors, they will drive prices down, so that level one competitors will have to spend a higher percentage of revenue on product design. If all competitors are level one (i.e., they are all inefficient), then prices may be kept artificially high and design costs to revenue kept artificially low. However, given all the drivers that we have discussed in Chapter 4, it is unlikely to stay that way for long.
3. The major PLM solutions providers and the major consulting firms have developed assessment methodologies that organizations can contract them to perform.

The Real World and the Universe of Possibilities for PLM

As with all major technological changes, PLM is not without its issues to be addressed—from technological barriers to social issues. This chapter will explore some of these major issues. On the technology side, the chapter will discuss issues such as federated spaces, information provenance, identity, and technology that connects real to virtual space. From the social side, the chapter will explore issues regarding intellectual property ownership, security, and privacy. Although it doesn't provide solutions, the chapter will point to the direction that future research and discussion should take.

Scenario I

You pull you car into the service bay as usual. However, instead of having the service representative greet you with the standard, "Hi, what can we do for you this morning?" She walks up to the front of your vehicle with a tablet PC for a moment. She then walks back to greet you with

Hi, Mr. Brown. I see you're here for your 50,000 mile service. it looks like there's a hose that's getting worn and needs to be replaced. We'll change your fuel mixture to optimize all of the highway driving that you're doing lately. We're going to replace the fuel pump. While it isn't under recall, the batch that part is from has caught the attention of the engineering guys, and they've asked us to replace them as a precaution. Also, one of the new tires you recently bought is a little low and is wearing a little funny. We'll look at it. Would you like us to bump out the quarter panel, where it looks like you had a brush with your garage door?"

Scenario 2

The senior design team for an aeronautical firm is standing around a holographic stage. Hovering above the stage is their latest fighter jet, which is currently in the design stage. Periodically, one of the engineers will call for some modification of the holographic image—for example, calling out a wingspan section and having that explode to show the various pieces and parts in it. Some of the components have privacy symbols on them. These are proprietary designs from component suppliers that cannot be examined nor exploded. However, engineers from these suppliers are present via a video wall and ready to answer questions about their respective parts.

The senior engineer asks for the plane to be put through its paces, and the group of engineers watch as it does acceleration tests, climb tests, roll tests, and other aerobatic maneuvers. At certain points, one or more of the engineers will stop and ask to see force readings from various aspects of the plane's surface. When an engineer from the team receives a reading that he doesn't like, he asks for a model of a previous plane to be brought onto the stage and color coded on the basis of failures. Green indicates that there is no failure with that component. The coloring goes all the way up to red, to indicate that there have been catastrophic failures. The part that he has a problem with in the new plane is coded red on the previous plane. The engineers immediately make comments, which are captured, about changing the design of that particular part.

Scenario 3

The Crime Scene Investigation Team (CSI) is called to the scene of a fatal hit-and-run-accident. A young man changing his tire by the side of the road was hit by a vehicle that fled the scene of the accident. The CSIs collect paint samples that they then analyze in the laboratory. They determine that the paint samples are from a certain make and model SUV. Using DMV records, they narrow the possible vehicles down to a specific SUV.

However, when they investigate the SUV, the CSIs find it undamaged. They then tap into the SUV's internal system. They discover that it recorded an impact on the right front bumper and alerted the On-Star-like system. The CSIs then contact the On-Star-like provider and download a digital recording of the operator contacting the driver, who answered in a slurred voice. The driver tells the operator that nothing happened. The slurred voice matches the voice of the owner of the SUV. The owner is arrested—betrayed by his SUV.[1]

In this last chapter, we will briefly attempt to reconcile the real world and the universe of possibilities for Product Lifecycle Management. The different scenarios above point to a direction that we think Product Lifecycle Management will take as it continues to evolve and mature. These scenarios are all real possibilities that PLM can bring about in the future.

The reason that they are plausible is twofold. First, none of the scenarios require any revolutionary technology breakthroughs. We're not assuming any new discoveries in faster-than-light transportation, fusion energy, or even nanotechnology. None of the events described above are outside the realm of possibility that today's technology can provide.

The second reason that these are plausible scenarios is the array of change that has occurred over the past 30 years. If we look at the distance that information technology has come during that period of time, these future possibilities might even look quaint. The difference between text reports printed 30 years ago on green-bar paper (so called because it alternated horizontal white and green bars on computer paper in order to break up the monotony of figures in page after page of reports), and today's natural-language inquiries against the worldwide database on Google is much greater than the difference between what we can do today and the scenarios at the beginning of the chapter. To understand the possibilities that simulation might hold for us, one needs only to compare the Pong games of yesteryear against the PS2 and X-Box offerings that exist today. It may not be possible to be too outlandish in imaging what might be possible in the next 30 years even if there is some slowing of exponential capability growth.

As we noted in Chapter 1, we're only on the twentieth square of the chessboard and the line of technology improvements with respect to processing power, storage, and bandwidth head straight north from here. As we discussed back in the first chapter, the doubling impact of Moore's Law will have far-reaching consequences for the capabilities that we will have with respect to information technology.

However, in keeping with our focus on PLM as an approach, the three aspects that we have concerned ourselves with—people, processes and practices, and technology—will also be a factor in deciding which scenarios are feasible in the future and which scenarios are not. In looking at the universe of possibilities, we will start with technology. When talking about the future, technology is always the first thing that people focus on as they assess the realm of possibilities.

Technology

One thing that we can be fairly sure of, given the exponential increase in processing, storage, and bandwidth, is that we will be able to handle more data, do more processing of it, and move it around in greater quantities than we have in the past. Given that we are in this doubling phase, the next wave of increases will double the capability of what we have done in the entirety of the past 30 years!

If we look at the Information Mirroring Model, which we reproduce here as Figure 11.1 the capabilities of virtual space on the right-hand side of the figure will only continue to increase. Virtual space will be able to reproduce what we experience in real space with increasingly better fidelity. Since we will be able to store more data, the representations of virtual objects will get richer and more complex. With increased processing power, we will be able to run more complex and more accurate simulations.

And we will run more of them and run them in shorter periods of time than we ever have in the past. Since we will do faster processing, computing types of uses—especially where the data is not well defined or we have to obtain data from a wide variety of sources—will become feasible in the near future. Meeting the Grieves Visual Test will be a given. Meeting the Grieves Performance Test will become commonplace. Computers may never be able to pass the Turing Test and imitate thinking human beings. However, they are well on their way to imitating everything else in the physical world.

There are still a number of challenges with working in virtual space. The three we will address briefly are federated spaces, information provenance, and identities.

The federated spaces concept is that of weaving individual shared spaces into a single logical space. Although our model shows the main virtual space as unique like its counterpart—real space—it is not. The reality is that we have a huge number of virtual spaces that are spread far and wide on the Internet. However, unlike our own personal mental spaces, we can share these virtual spaces and create logical views that act as one space. The ability to create this shared logical view out of different individual views, such that some information can be shared and other information protected, is what federated spaces will attempt to accomplish.

Since information is not tangible, it is difficult to determine the provenance or ownership of it through history. However, traceability

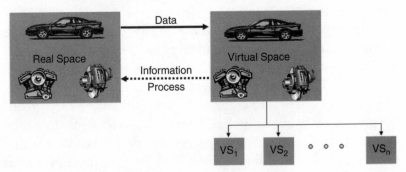

Figure 11.1   Information Mirroring Model

is a key element of PLM. For there to be traceability, there will need to be information provenance. We will need a mechanism to trace the source and history of information throughout its life.

At the beginning of the Internet era, there was a cartoon that showed a dog in front of a computer. The caption was, "On the Internet, Nobody Knows You're a Dog."[2] The problem turned out not to be dogs. In virtual space, individuals have no physical features, no identifying marks, no DNA to prove who they are. As we move more and more into virtual space, the ability to absolutely and unequivocally identify the human beings working there is one of the single biggest challenges of virtual space. While Internet identity fraud has received a great deal of attention, the general issue of the identity of the people who are working on products, submitting changes, and approving modifications is required in order to have both traceability and accountability.

However, the most interesting part of the Information Mirroring Model may be the linkages between real space and virtual space. It is where we expect we will see the most new and different uses of technology. The Information Mirroring Model relies on the following facts: we will be able to get the information we need about the changes in physical objects from real space and we will be able to transfer the information that we create in virtual space to real space on demand and with almost perfect fidelity.

The outcome is that we will see physical objects become more and more intelligent and conversant as we build more and more computers, sensors, and communication devices into our products. This means that products will be able to sense their own states of operation and the environment and be able to transmit changes in these to their image in virtual space.

The move to RFID, with each element of the product being able to identify itself, is already gathering momentum. Couple RFID with ubiquitous global positioning systems (GPS), and we have physical products that can sense changes in their environments, know exactly where they are in real space, and update their virtual doppelgangers continually in real time. The example that we used earlier of physically having to go out and find helicopters with a certain part will be a thing of the past. By interrogating virtual space, we will know where every single one of those helicopters is, what components are in the helicopters, and what state those components are in.

There will also be increasing requirements to transfer data and information from virtual space to physical space. There will be two recipients of this data and information: machines and human beings. Because products have increasing intelligence in them, even if only for sensing and reporting their status and environment, we will be able to make changes to their programs in virtual space and transfer those updates automatically to product-based computers.

This is a reflectivity of a different sort than we talked about in Chapter 3. But conceptually it is consistent, since it is the physical world that reflects the virtual world. Software will control more and more of the functionality of physical products and give us a wider range of possible functions that products will be able to perform without physical modification.

Going the other way, we as human beings will need better and better access to virtual space. Barring the development of hardware man/machine interfaces, which is not out of the realm of possibility, we are going to need to get better visual access to virtual space. Since our eyesight is our highest bandwidth input device, the requirement that we be able to see visually what we have in virtual space will be an absolute requirement. The ability of designers to visually inspect their designs and see the impact of a simulation is something that exists in relatively primitive form today but will get increasingly better.

We have moved rapidly from two-dimensional to three-dimensional views. It is only an implementation issue to be able to generate full-sized models holographically. In addition, we can change the parameters of the physical world in virtual space and visually be able to see things that before we could only measure or understand indirectly. Such things as air flows, stress vectors, heat gradients, and different substances can all be shown visually so that we can actually see the effect of a physical world that normally would be invisible to us.

With the rapid increases in wireless technology, the ability to access the virtual world at any place at any time and obtain the right information when we need it is not that far off. While we do not know the exact shape and form that these devices will take, they probably will develop along the lines of personal digital assistants (PDAs), although with much more capability and visual functionality than we have seen in the past. However these devices manifest themselves, it is pretty clear that the exciting work in technology will focus on this linkage and synchronization between the real and the virtual worlds.

People

While technology marches inexorably ahead, people scramble to catch up, make sense of it, and integrate it into their practices and processes. Product Lifecycle Management will be no exception. The social issues will either accelerate or constrain the adoption of PLM. The issue of who owns intellectual property in virtual space will become more and more crucial to resolve. In real space, the ownership issues are relatively simple. There is a unique physical object that we have possession of, or at least unique geospatial coordinates. Also, we generally can trace the history of where the product has been and who has had possession of it. Information is different because it can be so easily duplicated. Ownership is socially and legally constructed and has nothing to do with possession.

The example of the engineers standing around the holographic image of a fighter plane with certain pieces marked private is an indication that not everybody will be able to have access to every piece of information if we are going to protect intellectual ownership. Closely connected with the issue of ownership is the issue of security.

Controlling and/or preventing access to information is an area that is receiving a great deal of attention and substantial resources. This issue is one that has already arisen today with complaints about math-based designs being transferred illegally from one company to another.[3] This issue will continue to become more and more important.

The topic gets murkier when we add the requirements of collaboration to the mix. Collaboration, by its very definition, requires the sharing of information. This is why there are new areas of interest developing around such topics as federated spaces—which is using the information collaboratively but having the details reside in private, not shared space—and digital beaconing—the tracking and notification of the movement of math-based design information.[4]

Privacy is also an aspect that will have to be considered. In our first example, where we drive into the dealership, we may or may not want the dealer to know about aspects of what is undeniably our automobile—not the dealer's. For example, while it may be acceptable for the dealership to know that there are new tires on the car, for the dealer to know how much we paid for them from another supplier would be a breach of privacy. As Product Lifecycle Management evolves into collecting and assimilating in-use information in order to be able to understand the product as it is used and improve new versions of it, privacy issues will continue to be problematic.

However, as social systems evolve, so might this issue. One possibility is that the privacy issues are really a payment issue. It is not outside the realm of possibility that product manufacturers will pay for the information from the user of the product in order to improve their product. If payment were made to the product owner, this payment may be a way to alleviate the privacy issue for products. This is especially important where gaining access to the information involving the usage and performance of those products is critical in improving them.

Processes/Practices

We would also expect that the process/practice issue will continue to evolve. On the process side, the ability to turn over to a computer the analysis of more and more complex situations that previously required human judgment will continue to accelerate. PLM systems continue to add more and more experts and wizards into their products to provide assistance for design and analysis of increasingly sophisticated and complex components, products, and even entire systems. What once was the sole purview of human beings is rapidly being taken over by computers. This does not replace people; it frees them from routine analysis so they can focus on what computers cannot do—innovate. As computing capability continues its exponential increase, computers will take over tasks that were previously impossible for them to perform.

As mentioned earlier, the routing on the factory floor was once a practice performed by the factory supervisor because it was too complex for a computer. Today's computers can do that task easily. There will be other processes where complex activities that can now be done only by human beings will be performed by computer systems. In addition, as we evolve computer techniques such as fuzzy logic, we may be able to add the capability to process imprecise and ambiguous data elements, which we cannot do well using today's technology.

We also expect that computers will continue to assist in practices. The ability that allows people to search for exemplars or to detect patterns in information will be enhanced by the use of more and more powerful computers and their access to a wide range of data and information. The practice of medicine today is and will continue to be enhanced by having access to that information. If we can do that for biological systems, clearly the possibility for doing them for man-made manufactured systems is eminently possible.

This is only a very rudimentary and cursory look at the universe of possibilities for Product Lifecycle Management. Chances are that even the most esoteric scenarios that we can imagine have the possibility to become reality in a relatively short period of time. We will continue to be guided by our focus on Lean Thinking, which means substituting information for wasted time, energy, and material. And we will only continue to do those things that show up favorably when viewed from a financial perspective.

However, as technology increases on its exponential curve we would expect that things that were not feasible to do or too costly as looked at on the IT Value Map will become affordable and commonplace. Product Lifecycle Management, in a very short period of time, has made a tremendous impact in the way we look at products, the different functions to give those products, and the information that we both require and want as an asset for our future activities.

PLM is in its infancy, but has the promise of tremendous impact. PLM will drive the next generation of Lean Thinking across our organizations and into the supply chain by using product information as a replacement for wasted time, energy, and material.

Notes

1. This scenario was the subplot for an episode of the television series *CSI: Miami*. The episode was entitled, "Rap Sheet." The episode first aired on May 10, 2004.
2. P. Steiner (Artist), *On the Internet, Nobody Knows You're a Dog* [Cartoon], *The New Yorker*, July 5, 1993, p. 61.
3. See D. Roberts, "Did Spark Spark a Copycat?" *Business Week*, 64, February 7, 2005.
4. The leader in the area is Autoweb Communications, Inc, a subsidiary of NTT Communications, Inc., which is the largest clearinghouse for transferring and translating math-based designs between manufacturers and their supply chains. Autoweb is aggressively working on mechanisms to track and notify the owners of math-based design information when any movement of that information occurs. The author serves on its board of directors.

Index

Page numbers followed by *n.* indicate end-of-chapter notes.

About the Author

Dr. Michael Grieves founded the Product Lifecycle Management Development Consortium, University of Michigan's College of Engineering, and served as its Co-Director. He developed the first on-line Product Lifecycle Management Overview course for the College's Center for Professional Development and organizes and chairs the annual University of Michigan AUTO IT Conference. Grieves also is affiliated with the University of Arizona's internationally ranked MIS Department. Grieves works with PLM users and suppliers of companies such as General Electric, IBM, and Toyota on PLM strategies and implementations. With 35 years of industry experience, Grieves is a principal in the international management and IT consulting firm Core Strategies Inc., and serves on the boards of a number of technology companies.